Alternative Energy Systems
in Buildings

Alternative Energy Systems in Buildings

Special Issue Editors

Enrique Rosales Asensio
Antonio Colmenar Santos
David Borge Diez

MDPI • Basel • Beijing • Wuhan • Barcelona • Belgrade • Manchester • Tokyo • Cluj • Tianjin

Special Issue Editors

Enrique Rosales Asensio
University of La Laguna
Spain

Antonio Colmenar Santos
National University of Distance
Education
Spain

David Borge Diez
University of Léon
Spain

Editorial Office
MDPI
St. Alban-Anlage 66
4052 Basel, Switzerland

This is a reprint of articles from the Special Issue published online in the open access journal *Energies* (ISSN 1996-1073) (available at: https://www.mdpi.com/journal/energies/special_issues/building).

For citation purposes, cite each article independently as indicated on the article page online and as indicated below:

LastName, A.A.; LastName, B.B.; LastName, C.C. Article Title. *Journal Name* **Year**, *Article Number*, Page Range.

ISBN 978-3-03936-220-2 (Hbk)
ISBN 978-3-03936-221-9 (PDF)

Contents

About the Special Issue Editors

Enrique Rosales Asensio (Ph.D.) is an industrial engineer with postgraduate degrees in electrical engineering, business administration, and quality, health, safety and environment management systems. He has been a lecturer at the Department of Electrical, Systems and Control Engineering at the University of León, and a senior researcher at the University of La Laguna, where he has been involved in a water desalination project, in which the resulting surplus electricity and water were to be sold. He has also worked as a plant engineer for a company that focuses on the design, development and manufacture of waste-heat-recovery technology for large reciprocating engines, and as a project manager in a world-leading research centre. Currently, he is an associate professor at the Department of Electrical Engineering at the University of Las Palmas de Gran Canaria.

Antonio Colmenar Santos has been a senior lecturer in the field of Electrical Engineering at the Department of Electrical, Electronic and Control Engineering at the National Distance Education University (UNED), since June 2014. Dr. Colmenar-Santos was an adjunct lecturer at both the Department of Electronic Technology at the University of Alcalá and at the Department of Electric, Electronic and Control Engineering at UNED. He has also worked as a consultant for the INTECNA project (Nicaragua). He has been part of the Spanish section of the International Solar Energy Society (ISES), and of the Association for the Advancement of Computing in Education (AACE), working in a number of projects related to renewable energies and multimedia systems applied to teaching. He was the coordinator of both the virtualisation and telematic services at ETSII-UNED, and deputy head teacher and the head of the Department of Electrical, Electronics and Control Engineering at UNED. He is the author of more than 60 papers published in respected journals (http: //goo.gl/YqvYLk), and has participated in more than 100 national and international conferences.

David Borge Diez has a Ph.D. in Industrial Engineering and an M.Sc. in Industrial Engineering, both from the School of Industrial Engineering at the National Distance Education University (UNED). He is currently a lecturer and researcher at the Department of Electrical, Systems and Control Engineering at the University of León, Spain. He has been involved in many national and international research projects investigating energy efficiency and renewable energies. He has also worked in Spanish and international engineering companies in the field of energy efficiency and renewable energy for over eight years. He has authored more than 40 publications in international peer-reviewed research journals and participated in numerous international conferences.

Preface to "Alternative Energy Systems in Buildings"

Energy conservation through energy efficiency in buildings has acquired prime importance all over the world. The four main aspects of energy efficiency in a building include, first and foremost, the nearly zero energy passive building design before actual construction; secondly, the usage of low energy building materials during its construction; thirdly, the use of energy efficient equipments for low operational energy requirement; and lastly, the integration of renewable energy technologies for various applications. This Special Issue, published in the Energies journal, includes five contributions from across the world, including a wide range of applications, such as desiccant-assisted air conditioning systems, novel prototype design for building-integrated wind turbines, photovoltaic–wind hybrid plants integrated into an urban environment, and solar energy harvesting technologies for building integration and distributed energy generation. Finally, we wish to express our deep gratitude to all the authors and reviewers who have significantly contributed to this Special Issue. Our sincere thanks also go to the editorial team of MDPI and Energies for giving us the opportunity to publish this book, and for helping in all possible ways, especially Ms Wu, for her precious support and availability.

Enrique Rosales Asensio, Antonio Colmenar Santos, David Borge Diez
Special Issue Editors

Review

Recent Developments in Solar Energy-Harvesting Technologies for Building Integration and Distributed Energy Generation

Mikhail Vasiliev *, Mohammad Nur-E-Alam and Kamal Alameh

Electron Science Research Institute (ESRI), School of Science, Edith Cowan University, 270 Joondalup Dr, Joondalup, WA 6027, Australia; m.nur-e-alam@ecu.edu.au (M.N.-E.-A.); k.alameh@ecu.edu.au (K.A.)
* Correspondence: m.vasiliev@ecu.edu.au

Received: 14 February 2019; Accepted: 15 March 2019; Published: 20 March 2019

Abstract: We present a review of the current state of the field for a rapidly evolving group of technologies related to solar energy harvesting in built environments. In particular, we focus on recent achievements in enabling the widespread distributed generation of electric energy assisted by energy capture in semi-transparent or even optically clear glazing systems and building wall areas. Whilst concentrating on recent cutting-edge results achieved in the integration of traditional photovoltaic device types into novel concentrator-type windows and glazings, we compare the main performance characteristics reported with these using more conventional (opaque or semi-transparent) solar cell technologies. A critical overview of the current status and future application potential of multiple existing and emergent energy harvesting technologies for building integration is provided.

Keywords: renewables; energy saving and generation; built environments; transparent concentrators; luminescent concentrators; solar windows; advanced glazings; photovoltaics

---

1. Introduction

Worldwide annual energy consumption is projected to exceed 700 quadrillion British thermal units (Btu), or 0.74 billion TJ by 2040, with the energy generation contributions from fuels other than coal (mainly renewables) being on the increase currently [1]. Around 22.7 billion tons of anthracite coal fuel is needed to release the thermal energy equivalent of to this annual energy consumption figure. At the same time, the combustion of fossil fuels remains among the main concerns identified in relation to the past and current global warming and environmental pollution trends [2–4]. Considering this, the development of various types of energy-saving approaches and novel energy generation technologies is of increasing importance today, especially in the building and construction sectors where a substantial fraction of the total energy generated worldwide is being used. In the US and the EU, buildings now account for over 40% of the total energy consumption [5]. At present, the technologies for on-site distributed renewable energy generation in built environments are experiencing rapid advances, yet their widespread utilization is still some years away from being commonplace with one exception, the ubiquitous deployment of conventional photovoltaics (PV) on residential building roofs. Building-integrated PV (BIPV) technologies, in a variety of possible implementations, are widely expected to play a large (and growing) role in near-future construction practices, complementing the now-mature energy-saving construction technologies. A recent report by the European Commission [6] specifies a new societal mission that could be called "creating the Internet of Electricity." This new term means achieving the fundamental transformation of the power system based on widespread and distributed use of renewables, integrating energy storage, transmission, dispatchment through the smart use of energy consumption. This "Internet of Electricity" is now seen as a fundamental step towards the full integration and decarbonisation of the entire energy system.

Energy-efficient buildings, construction materials, windows, and vehicles are gaining significant attention and increasing importance today [7–13]. The on-site generation of renewable energy coupled with using energy-efficient construction materials and energy-saving appliances forms a viable, future-proof approach to building the infrastructure and vehicles of tomorrow, in practically all geographic regions. The concept of a zero-energy building (ZEB) was first mentioned in 2000 and became a mainstream idea by 2006 [14]. Technologies for enabling widespread heating and also cooling-related energy savings in buildings through reducing the thermal emittance of glass surfaces have a much longer history, dating back to at least the early 1970's [15,16]. Since then, a large number of research works have been dedicated to achieving continually improved control over the various performance aspects of modern energy-efficient coatings and window glazings, such as their visible-range light transmission (VLT), solar heat gain coefficient (SHGC), thermal insulation performance (U-value), and the ability to control window tint actively or passively. Excellent reviews of key developments in these areas are now available [17,18]. Among the more novel, recently-developed approaches to preventing the overheating of building surfaces are the use of coatings for "passive radiative cooling," which force the re-emission of the absorbed thermal energy within the atmospheric infrared transparency window between 8 and 13 μm, thus, utilizing the vacuum of space as a heat sink [10,11]. In recent years, the now-traditional spectrally-selective metal-dielectric low-emissivity coatings have found an additional niche application area, serving as components of novel energy-harvesting photovoltaic solar windows [19–21], whilst maintaining their energy-saving functionality at the same time. Multiple BIPV-based solar and solar-thermal energy harvesting approaches now form the foundations for a diverse group of mature, industry-ready technologies, with their application areas and markets growing rapidly [22–25]. At the same time, most semi-transparent, and especially highly-transparent BIPV product types, are only beginning to fill their potentially widespread, yet still, niche-type, application areas, and are at present widely considered as "disruptive technologies", due to their relatively short history of development and commercialisation [26].

Comprehensive reports and reviews on the types of modern BIPV installations, their economics, performance, and current industry trends are available from [27–30]. Large-scale installations of semi-transparent BIPV module types in building facades still remain much rarer than conventional BIPV roofs, canopies, façades, and wall coverages, whether colour-adjusted or conventional. Figure 1 provides a graphical summary of the broad range of the building-applied PV (BAPV) and also BIPV technologies, materials, modules, and application types, which are either in common use at present, or beginning to appear on the market.

Energy generation (or energy harvesting) has not traditionally been associated with building walls, windows, or any glazing products, until (perhaps) the current decade. Various approaches to the incorporation of photovoltaic (PV) systems into building envelopes began being actively explored in the last several years, leading to significant growth in this new field of building-integrated photovoltaics. Historically, the building-integrated solar energy harvesting installations started as façade- and wall-integrated conventional (Si, CdTe, or CuIn(Ga)Se$_2$) PV modules occupying the building envelope areas other than roof surfaces, and continued towards the development of semi-transparent, glass-integrated PV window systems using patterned amorphous-silicon modules, perovskite-based, or dye-sensitised solar cells, e.g., [26,30–35].

More recently, the field of luminescent solar concentrators (LSCs, [36–38]) has begun showing clear signs of a "renaissance" [39–42]. There has also been significant progress demonstrated in the development of semi-transparent organic, polymer-type, and also perovskite-based solar cells, and organic materials-based transparent LSC [43–46], driven by the opportunities to capture the growing markets in both distributed electricity generation and advanced construction [6,19,20,31]. Up to date, practically no installation-ready solar windows using transparent organic solar cells or LSC have been marketed as industry standards-compliant building material products.

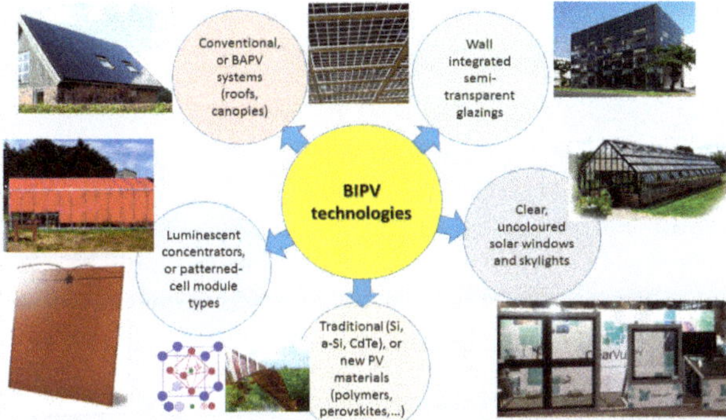

Figure 1. Building-integrated photovoltaic (BIPV) modules, technologies, applications, and materials—conventional and emergent.

Windows and glazing systems, despite being a practically ancient technology at its core, are now starting to be viewed and recognised as being the renewable energy platform and resource of the near future [26,47]. This potential for enabling extremely widespread energy harvesting at the point of use is practically unparalleled, considering the entire history of industrial energy generation, and is expected to develop hand-in-hand with the growing use of transparent heat-regulating (THR) coatings for energy-saving applications. Expanding the worldwide energy generation facilities into the almost-uncharted, yet vast territory of glass and windows will require significant time, research efforts, broad-based investments, and long-term strategic thinking on behalf of governments and private companies. This is due to the fundamental and crucial ways in which the energy industry differs from all other industries, which has been underscored in a recent publication [48] by the Breakthrough Energy Coalition, authored by B. Gates. The field of distributed, building-based energy generation and the era of the Internet of Electricity are both in their infancy today, with major new developments and discoveries still waiting to happen. Despite the intrinsic challenge of generating appreciable amounts of electric energy from sunlight using energy-capturing components which themselves require substantially visible transparency, significant progress has been demonstrated in BIPV technologies in recent years. The main aim of this present work is to highlight the important recent developments in the approaches, materials, structures, and systems dedicated to making widely distributed renewable energy generation in built environments a reality. The next sections of this article are structured to first describe (within Section 2) a wide range of technologies available currently for enabling solar energy capture from building surfaces, and their main metrics parameters; other sections are dedicated to describing the principal development milestones relevant to next-generation BIPV achieved recently in both research labs and in industry. Development progress over the last decade in two different classes of BIPV systems—the semitransparent non-concentrating solar modules, and in semitransparent solar concentrators—is reviewed, clearly separating the results achieved in research samples from product-level datasets. Main concentrator-type solar window metrics parameters and the physical effects limiting the achievable performance characteristics are described in Section 3.2, together with materials-related considerations. A discussion of how the system limitations are being addressed by different research groups is also provided. Section 3.3 provides a description of sample applications of transparent solar windows, showcasing an immediate application area of the harvested energy at the point of generation—inside the window structure itself where active control over transparency is possible. Other application areas are also mentioned, together with a brief discussion of significant recent achievements in semitransparent organics-based solar cell materials. This review focuses mainly on the developments in three principal technology categories (regular BIPV, non-concentrating semi-transparent BIPV, and LSC-type devices), illustrating the recent history and evolution of unconventional photovoltaics, developing towards systems with increasing power conversion efficiency simultaneously with improving control over the system appearance, architectural deployment suitability, product lifetime, and application readiness.

2. Main Technologies for Integrating Energy Harvesting Surfaces into Buildings

Expansion of the potential deployment areas for the traditional (non-transparent) PV modules started with the use of building façade and wall surfaces. This was likely due to both the ready availability of these additional energy-harvesting areas, and also because of the relative scarcity of the optimally-tilted roof-based PV placement options, especially in multi-storey urban environments. Additional efforts aimed at further expanding the surface areas suitable for "solarisation" included the placement of PV modules in somewhat unexpected locations, e.g., under road pavements [49]. Even though both the horizontal and the vertical module orientations are not optimally tilted with respect to the incoming sunlight, vast potential deployment areas become available using these approaches. The placement geometry of these non-conventional energy harvesting surfaces can be customised in a site-specific manner to maximise the energy production efficiency in most locations, by accounting for local environment-specific variables, such as the prevailing sun azimuth direction during the summer months and external shading conditions. Considering the sun altitude angle corresponding to the standardised peak irradiation conditions (AM1.5G spectral distribution at 1000 W/m^2), both the horizontal and the sun-facing vertical PV surfaces intercept about 700 W of the total (direct-beam and diffused) solar irradiation flux per 1 m^2 of active area at peak weather conditions. The azimuth-optimised vertical placement of PV surfaces can be more suitable for maximizing the yearly energy output per building footprint area (compared to the horizontal orientation), at least for urban locations in moderate latitudes. This is because of factors, such as the accumulation rates of surface contaminants, wind-assisted cooling effects, sun altitude angles being well away from zenith for most of the day, and the ground albedo or building-wall reflections, which provide an additional diffused radiation background easily interceptible by the wall-mounted PV. At the same time, the overall architectural design of buildings should ideally account for the site-specific and climate-specific energy-harvesting performance optimisation of wall-mounted PV arrays or windows, for example, by installing these systems on one or two of the most suitable building walls only. Figure 2 provides a system-level graphical outlook and main performance comparisons for most of the BIPV technology types commercialised so far.

Figure 2. Conventional (building-applied photovoltaics (BAPV)), colour-optimised, and semitransparent commercially available BIPV technologies at a glance. (**a**) Avancis PowerMax Skala CuInSe$_2$ panels [50]; (**b**) Multilayer-coated, colour-optimised BIPV facade by EPFL (Ecole Polytechnique Federale de Lausanne, Switzerland) and Emirates Insolaire [27,28,33]; (**c**) AGC (Asahi Glass Corporation, Japan) Sunjoule product [51]; (**d**) Onyx Solar a-Si high-transparency BIPV panels [52]; (**e**) Hanergy BIPV panels using a-Si [53]; (**f**) High-transparency CdTe BIPV panels [35]; (**g**) Solaronix BIPV façade based on semi-transparent dye-sensitised solar cells [34,54]; the methodology used for making the estimates of electric output is described in [55].

The average transparency-related and energy-related figures of performance shown within insets in Figure 2 have either been estimated from the published data, or taken from the relevant product specifications. The standardised peak-rated electric power outputs per unit active PV area, shown as P_{max} or W_p/m^2 data within parts of Figure 2, have been obtained from the published manufacturer's specifications, in which the optimum (peak-output) geometric orientations and tilt angles were presumed, except for Figure 2a. The P_{max} figure shown in Figure 2a was obtained from Avancis, Inc. published product specifications by also accounting for the vertical sun-facing panel orientation, using the flux reduction factor of 0.7. The estimated figures for the electric output per unit active area of custom-installed BIPV (Figure 2b,c,g) have been obtained using the published data for the yearly energy outputs, the total areas of installed PV, and the location-specific weather-dependent insolation data, using the methodology described in [55]. Therefore, these estimates of the maximum expected electric power output per unit active area are not standardised with respect to either the incident solar spectrum or cell surface temperatures.

A notable recent trend in BIPV has been the apparent "mimicry" capability of the solar cell surfaces covering building facades, assisted by the reflection colour-tuning multilayer thin-film coatings. Twelve thousand coloured solar panels have been installed at the Copenhagen International School's new building (Figure 2b), completely covering the building and providing it with 300 MWh of electricity per year (and meeting over half of the school's energy needs) [27,33]. These PV panels covered a total area of 6048 square meters, making it one of the largest BIPV installations in Denmark [28]. It is possible to derive a figure of performance of about 62 W/m^2 for the maximum expected electric output generation capacity, by using these reported data on the predicted annual energy generation, the energy-converting area installed within the façade (6048 m^2), and by approximating the other parameters (e.g., assuming the peak-equivalent sunshine-hours per sunny day at the installation location is around 4 h, and 200 sunny days per year). The annual number of sunny days is approximated here by using the figures from the average monthly distribution of rainy days for this location, which is accessible from a range of online weather-related data sources. The multilayer coatings, which provided the apparent colour adjustment by reflecting the blue-green parts of the spectrum, have, therefore, reduced the electric performance somewhat, compared to an optimally-oriented $CuInSe_2$ (CIS) facade. However, this 62 W/m^2 figure has been obtained from a real, feature-rich architectural installation in Northern Europe, in which a significant fraction of active PV area has not been oriented optimally, and also experiences partial geometric shading. This shows the significant practical application potential of colour-adjusted BIPV technologies, at least, for the non-transparent installations. Similar performance in energy harvesting (~58 W/m^2) has been estimated from a horizontally-mounted semitransparent BIPV using monocrystalline silicon cell technology ([51], Figure 2c), as well as documented in [35] (~60 W_p/m^2) for a peak-oriented CdTe-based semitransparent (T_{vis} ~ 33%) non-concentrating BIPV module, likely from the product range of Xiamen Solar First Energy Technology Co., Ltd. (Xiamen, China)—judging by the close matching of the academically- and commercially-published electrical specifications ([35] vs. [56]).

It is interesting to compare the current energy-harvesting performance in the available semitransparent BIPV products with both the PV efficiency records achieved so far in small-size luminescent concentrators, and the theoretical limits of efficiency predicted for the highly-transparent concentrator-type BIPV, and also the transparent organic solar-cell modules. The current efficiency record for a 5 cm × 5 cm LSC using organic luminophores and edge-mounted GaAs cells stands at 7.1% [57], corresponding theoretically to 71 W_p/m^2. However, the scaling of electric power output cannot be linear with increasing concentrator area, for multiple reasons including the relevant loss mechanisms [58], and other considerations related to the thermodynamics of light concentration and light transport phenomena, discussed in subsequent sections. The assessments of the performance limits in highly transparent area-distributed PV and also in concentrators have been made in [59] and in [55], pointing to a theoretical possibility of generating up to about 57 W_p/m^2 in systems of 70% colour-unbiased transparency. This theory-limit performance was calculated presuming the use of

CIS solar cells of wide spectral responsivity bandwidth, at 25 °C cell temperature, 12.2% PV module efficiency, for the peak geometric orientation and tilt of idealised concentrator panels (with AM1.5G, 1000 W/m^2 irradiation.) The practically-achieved, literature-reported clear solar window performance in factory-assembled glass-based windows is now close to 50% of its theoretical limit [55]. At the same time, the best power conversion efficiency (PCE) reported recently in transparent organic photovoltaics was 9.77% at 32% transparency, according to [60]. The authors of [61] also reported achieving 4.00% PCE at 64% transparency in polymer solar cells produced by solution processing; other recently-demonstrated combinations of PCE and visible-range transparency in organic solar cells were summarised recently in [44]. To the best of our knowledge, no installation-ready or standards-compliant BIPV systems with academically published specifications and using transparent polymers-based solar cells (or transparent organic luminophores) are currently available on the market. Significant and ongoing product development efforts are being undertaken at Ubiquitous Energy (USA), aimed at commercialisation of transparent organics-based solar windows, with some groundbreaking material development results reported within supplemetary material dataset of [44], e.g., achieving PCE of 5.20% at T_{vis} = 52%.

Reports on the ready availability of any inorganic materials-based clear and highly transparent solar window, skylight, or curtain wall products are still very rare. Among these product types with published specifications and now available to the market are BAPV-type solar-powered skylights from Velux (Denmark) [62], and an emergent range of solar windows, curtain wall, and solar skylight products marketed by ClearVue Technologies (Perth, Australia), which have passed the various industry standard compliance tests in 2018. The relevant technical details and the performance-related description of ClearVue solar window prototypes are available from [55], and their core technology fundamentals and the history of development were reported in [19] and [20]. Other solar window manufacturers, e.g., Physee (The Netherlands) [63], or GlassToPower [64] do not appear to publish the technical details (in particular, PV current-voltage (I-V) curve datasets) related to their current product specifications.

Even though we are still at the very beginnings of the era marked by the widespread use of transparent (or clear) solar windows, the range and scale of their potential applications is recognised as enormous (summarised graphically in Figure 3.) The necessity of developing the new, windows-based distributed generation networks to future-proof the urban areas, power the Internet of Things (IOT) revolution, and reduce the reliance on fossil fuels has also been widely recognised [26]. This is further confirmed by the ongoing research, development, and investment momentum now continuing in this area and all related materials science areas worldwide [30–32,39–48].

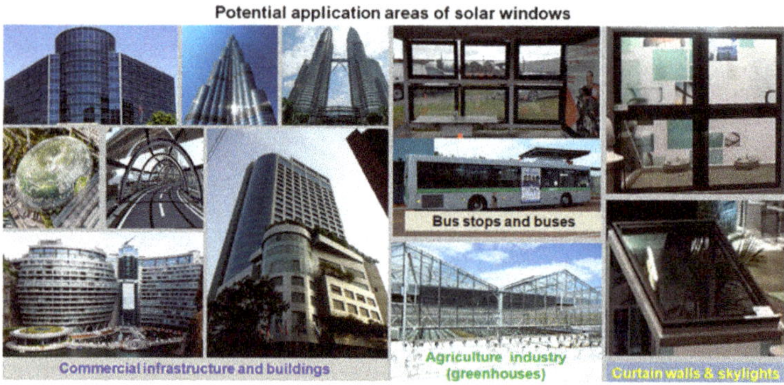

Figure 3. Highly transparent solar windows and their application areas. The solar window prototypes shown installed into an off-grid bus stop in Melbourne (Australia) are described in [55]; other solar windows shown in the right-hand side of the image are current products from ClearVue Technologies, showcased at Greenbuild Expo in Chicago, USA, in November 2018.

The value of developing highly-transparent solar windows is related to multiple unique qualities these systems can bring about. Among these are the provision of high-quality views and natural daylighting options for building occupants, the potential for large reductions in lighting-related energy expenditures, and the optional ready availability of added active or passive control over the window features, such as apparent colours or the degree of visible transparency. These features will require adding custom-designed optical coatings or active transparency-control layers to the initially-transparent energy-generating window systems, to maximise the number of possible options for the product appearance modification. Other unique benefits of highly transparent solar windows will be best illustrated in emergent application areas, such as advanced sustainable greenhousing, where the plant growth processes require either plenty of natural visible light or the precise control over illumination spectra.

Due to the renewed attention to the next-generation photovoltaics now being paid by multiple research groups, public institutions, and private companies worldwide, it is currently widely expected that new types of technologies, functional materials, and products will continue to be developed. The next sections of the present review will focus in more technical detail on the major results and developments demonstrated in recent years in the areas related to both the direct area-based solar energy converters, and also the concentrator-type solar windows.

3. Principal Results in Semi-Transparent PV Module Development, New Materials for Solar Concentrators, and Current Trends in Transparent Energy Harvesters

Considering that a number of important developments have been demonstrated in recent years in all BIPV technologies, related functional materials, and solar energy harvesting approaches, it is logical to identify the two main technology-related device categories to be discussed separately: the area-based PV energy converters, and the concentrator-type PV energy-harvesting systems. Forward-looking technologies and advanced novel materials have been investigated actively during the recent decade, resulting in noteworthy prototype demonstrations and product-level systems development, across the entire spectrum of BIPV application types. The following subsections address the principal results demonstrated in the area of semi-transparent PV energy harvesters from all main technological categories.

3.1. Recent Developments in Semitransparent Non-Concentrating BIPV Technologies

Semitransparent non-concentrating BIPV technologies are defined not only by their degree of visible-range transparency but also by their capability of enabling immediate photovoltaic energy conversion process, localised at any arbitrary region of light-ray incidence onto their active device areas. On the other hand, in concentrator-type semitransparent PV, the PV conversion is engineered to typically take place at the active (non-transparent) solar-cell areas placed at (or near) the edges of light-capturing semitransparent or clear aperture areas. These aperture areas serve to re-direct the incident light rays towards PV cells, and may contain materials providing partial light-trapping functionality, and/or light-harvesting structures that assist waveguiding-type propagation. Therefore, the major defining difference between these two main device categories is in the light propagation path-length within the devices, between the points of ray incidence and the location(s) where the PV conversion takes place. Both technological approaches possess their unique advantages and disadvantages, which are related to how the fundamental problem of balancing the overall energy-conversion efficiency and the degree of device transparency is addressed. The common metrics-related parameters used to evaluate the performance of semitransparent non-concentrating PV (or their suitability for any particular application area) include the visible light transmission (VLT), and power conversion efficiency (PCE) at standard test conditions (STC). The STC in PV metrology refer to using the standardised solar spectrum for device irradiation (AM1.5G spectral distribution, with 1000 W/m^2 in radiation flux density), and making PV current-voltage characteristic measurements whilst keeping the solar cell surface temperatures at 25 °C. If detailed PCE characterisation metrology results are not available, other published product specifications or performance-related data can often be used to derive or estimate the maximum output power rating per unit active area. Table 1 summarises the main performance parameters achieved using non-concentrating semitransparent PV technologies in recent years.

Table 1. Performance summary (transparency, materials, and efficiency-related data) for main non-concentrating semi-transparent solar cell technologies and building-integrated photovoltaics (BIPV) products with significant visible transmission.

Technology	Ref./Year	R&D Sample or Product	VLT	PCE or P_{max} (est.)	Materials/Details
Dye-sensitised solar cells	[65]/2007	sample	~60%@ 550 nm	9.2%	Screen-printed TiO_2 films
Dye-sensitised solar cells (Solaronix)	[34,54]/2014	product	N/A	~28 W/m², vert.	Evaluated from the available published data
Transparent PV solar cells	[66]/2011	sample	>65%	(1.3±0.1)	Organic material-based, harvesting near-IR only
Transparent polymer SC	[61]/2012	sample	64%	4.00%	Solution processing technology
Semi-transparent organic SC	[60]/2017	sample	32%	9.77%	Organics (dithienocyclopentathieno [3,2-*b*]thiophene)
Perovskite SC	[67]/2015	sample	~77%@ 800 nm peak	11.71%	Semi-transparent MAPbI$_3$ cell with Ag-nanowire transparent electrode; for use in tandem cells.
Single-junction semitransparent perovskite SC	[68]/2014	sample	1) 29% 2) 22%	1) 6.4% 2) 7.3%	Methylammonium lead iodide perovskite ($CH_3NH_3PbI_3$)
Colloidal Quantum Dot SC	[69]/2016	sample	24.1% (ave.)	5.4%	PbS colloidal QDs
BAPV glass-integrated PV roof (AGC Sunjoule)	[51]/2015	product	~20% of clear glass area	~58 W/m², horiz.	Mono-Si cells, separated laterally within glass
Hanergy BIPV panels	[53]/2018	product	40% (ave.)	3.8%	Amorphous silicon
Onyx Solar BIPV panels	[52]/2018	product	30% (ave.)	2.8%	Amorphous silicon
Solar First Energy Technology Co., Ltd.	[35]/2018	product	~33%	6%	CdTe semitransparent BIPV modules
Polysolar BIPV	[70]/2018	product	50% (ave.)	~55.5 W_p/m² (5.55%)	CdTe PS-CT-40 BIPV modules (1200 × 600 × 7 mm)
Stability-enhanced perovskite SC	[71]/2018	sample	N/A	Up to 20.2%	SnO$_2$ electron transport layer replacing TiO$_2$. T$_{80}$ operational lifetime of 625 h.

[1] VLT: visible light transmission, either spectrally averaged or related to a transmission peak at a specified wavelength. PCE: power conversion efficiency; P_{max} is rated (or estimated) maximum electric output power per unit active module area.

A number of industrialised, product-level semitransparent PV module manufacturing technologies have been established, based on patterning the non-transparent active PV material area to ensure transparency. Systems using either amorphous silicon or cadmium telluride are becoming increasingly common, even though their (spectrally averaged) visible-range light transmission does not exceed 40 to 50%. Up to date, none of the commercially available patterned-layer energy-generating modules employing inorganic PV materials feature colour-neutral clear appearance or uniform transmission characteristics across their aperture areas.

A notable current trend in semitransparent PV modules development is the continued (and growing) attention of multiple research groups worldwide dedicated to optimizing the perovskite-based material systems. This is due to the rapid progress demonstrated in the performance (PCE) of perovskite-based photovoltaics recently, and over a relatively short time scale, reaching

27.3% conversion efficiency record in (non-transparent) tandem-type perovskite-silicon solar cells in 2018 [72]. The recently-achieved record efficiency in perovskite-based solar cells is at 20.9% [73]. The PCE of other semi-transparent perovskite-based research samples is currently in excess of 20%, demonstrated in stability-enhanced systems, in which the titanium dioxide-based electron transport layer was replaced by tin oxide [71]. The operational lifetime of these perovskite-based cells reported at the end of 2018 is still on the scale of several months, rather than years, which places their commercialisation prospects within the scope of near-future years, rather than today. Due to the potentially vast future application areas of perovskite-based semitransparent BIPV, significant research efforts are continuing in the materials science areas related to optimizing the chemistry and module structure of perovskite-based devices, to ensure improved environmental stability and performance. Methylammonium lead iodide (MAPbI$_3$) based perovskite systems were recently shown [74] to possess intrinsic environmental stability limitations due to chemical chain reactions involving iodine vapour. The authors of [74] also hold a view that it is imperative to develop other types of perovskite material systems to achieve long-term stable solar cells, and work in this area is also ongoing in multiple research groups. Notably, efforts aimed at reducing the perovskite toxicity caused by the presence of lead and also removing the hysteresis in current-voltage characteristics have been reported [75,76].

Another rapidly developing category of semitransparent BIPV technologies possessing strong potential for future applications uses polymer-based organic materials, which can provide remarkable visible transparency simultaneously with high power conversion efficiency [60,61,66]. In this area, efforts to up-scale the solar cell sample size and ensure the long-term environmentally stable operation and commercially-applicable system packaging are currently ongoing, with product-level devices widely expected to be demonstrated in the near future.

3.2. Progress in Semitransparent Concentrator-Type Solar Window Technologies

Most semitransparent concentrator-type solar window technologies developed so far rely on the continued advances in the field of luminescent solar concentrators, which are underpinned by the development of new luminescent material types. This is because the only long-range (in the absence of glass surface imperfections or strong scattering) photon transport mechanism suitable for trapping the incident light energy within transparent waveguides is total internal reflection (TIR), which itself is enabled by the random directional character of luminescent emissions. The internal structure of LSC-type devices has also undergone rapid development, relying on the advances in areas, such as application-specific thin-film coatings, spectrally-selective transparent diffractive optics [20,77], embedded Mie scattering media [19,78], or other components designed to stimulate partial light trapping within waveguide-type glazing systems. More recently, conventional (non-transparent) PV elements, e.g., silicon or copper indium-gallium selenide (CIGS) cell modules started to merge into the design structure of transparent window-type solar concentrators, blocking a small transparent area fraction, but boosting the electric output through both the direct incident light capture and also collecting a part of light travelling within the device [40,55,79–81]. Various performance metrics parameters have been introduced in the LSC field for characterizing the performance of concentrator-type PV devices of all configurations and types; the most important of which are the PCE, geometric gain (G)—the ratio of the total light collection aperture area to the area of edge-mounted PV cells, photon collection probability P represented by the ratio of device PCE to the nominal PCE of the solar cell modules used in the device, and the optical concentration factor $C_{opt} = G \cdot P$ [82]. Another useful concentrator metrics parameter is optical power efficiency η_{opt} which is the ratio of the optical power received at the PV cell surfaces to the total optical power incident onto device aperture area. Table 2 summarises the main optical and electric performance metrics parameters demonstrated in either laboratory samples or product-level technologies using different concentrator-type semitransparent PV systems in the last decade.

Table 2. Performance summary for main recently-reported concentrator-type semi-transparent luminescent solar concentrator (LSC) devices and solar window technologies.

Technology	Ref./Year	R&D Sample or Product	VLT	PCE or P_{max}	Concentrator Performance Parameters	Materials/Details
5 cm × 5 cm LSC, non-transparent (using >97% diffuse backside reflector)	[57]/2008	sample	<3%	7.1%	G = 2.5; P ≈ 0.246; C_{opt} ≈ 0.616	Lumogen F Red305 and Fluorescence Yellow CRS040; 4 GaAs cells at edges. Highest LSC efficiency to date.
Quantum-dot LSC, 10 cm × 10 cm	[42]/2018	sample	43.7%	2.18%	P ≈ 0.198; η_{opt} = 8.1%	$CuInS_2$/ZnS QDs and poly-Si cells (η_{Si} = 11%); PCE = 2.94%, if backside reflector is used.
Tandem Quantum-dot LSC, 15.2 cm × 15.2 cm	[83]/2018	sample	~30%	3.1%	P ≈ 0.11;	$CuInSe_2$/ZnS QDs; high-efficiency GaAs cells.
Quantum-dot LSC, 12 cm × 12 cm	[84]/2017	sample	70%	N/A	η_{opt} = 2.85%	Si QDs.
50 cm × 50cm × 6cm LSC	[85]/2015	sample	est. ~40% at peak	1.26%	G = 20.83; P ≈ 0.057; C_{opt} ≈ 1.187	Sc-Si (22% eff.) cells at edges. Organic dyes (DTB, DPA).
163 cm × 63 cm LSC panels for greenhouse applications	[80]/2016	product	N/A (red coloured)	3.4%	P ≈ 0.17;	Mono-Si cells (20%eff.) at back surface (13.9% area coverage, straight lines pattern). Lumogen Red 305.
163 cm × 63 cm LSC panels for greenhouse applications	[80]/2016	product	N/A (red coloured)	3.8%	P ≈ 0.19;	Mono-Si cells (20%eff.) at back surface (13.9% area coverage, criss-crossed pattern). Lumogen Red 305.
"Leaf Roof"-type 110 cm × 0.5 cm LSC	[81]/2017	sample	N/A (multi-coloured)	5.8%	P ≈ 0.258;	Si cells (22.5% eff.) Multiple backside cells; Lumogen Red 305 mixed with other Lumogen pigments.
100 cm × 75 cm clear glass windows of hybrid concentrator type	[55]/2018	product	>65% (T_{direct}) ≈ 70% (T_{total})	~25W_p/m^2 $\eta_{conc.}$ ≈ 1.425% η_{total} ≈ 2.5%	G = 4.25; P ≈ 0.116; C_{opt} ≈ 0.492	$CuInSe_2$ cell modules (12.3% eff.); inorganic luminescent phosphor pigments; transparent near-IR reflector coating at back surface.
20 cm × 20 cm semitransparent diffraction-assisted LSC	[77]/2018	sample	>60% (T_{direct})	≈2.347%	G = 2.8; P ≈ 0.176; C_{opt} ≈ 0.495	$CuInSe_2$ cell modules (13.3% eff.) at edges; inorganic luminescent phosphor pigments; transparent near-IR reflector coating at back surface; embedded large-area transparent diffractive element.

[1] VLT: visible light transmission, either spectrally averaged or related to a transmission peak at a specified wavelength. PCE: power conversion efficiency; P_{max} is rated (or estimated) maximum electric output power per unit active module area. G: geometric gain; P is photon collection probability, and C_{opt} is optical power concentration factor; detailed definitions for these quantities are available from [82]. The figure for $\eta_{conc.}$ from [55] has been evaluated accounting for the concentrator-related contribution to the total system PCE of near 2.5%.

The optical power efficiency of LSC, or other types of concentrating semi-transparent PV energy harvesters (which utilise physical mechanisms other than TIR-guided luminescence, e.g., diffractive optics), is itself a product of multiple efficiency factors, each relating to a particular physical process harnessed for re-routing the incident photons towards the solar cell surfaces. For "classical" luminescent concentrator systems, η_{opt} can be represented in terms of the contributions of all physical phenomena taking place within the concentrator volume, in the following way [37,38]:

$$\eta_{opt} = (1 - R) \cdot P_{TIR} \cdot \eta_{abs} \cdot \eta_{PLQY} \cdot \eta_{Stokes} \cdot \eta_{host} \cdot \eta_{TIR} \cdot \eta_{self}, \tag{1}$$

where R is the reflectivity of the front surface of the luminescent waveguide; P_{TIR} is the probability of total internal reflection governed by the difference between the refractive indices of waveguiding material and outside air; η_{abs} is the fraction of the incident solar energy absorbed by the luminophore(s); η_{PLQY} is the photoluminescence quantum yield of the luminophore(s) used; η_{Stokes} is the efficiency factor characterizing the energy losses due to heat generation during the absorption and emission events (Stokes shift loss); η_{host} is the transport efficiency of the waveguide; η_{TIR} is the reflection efficiency of the waveguide determined by the smoothness of the waveguide surface, and η_{self} is the transport efficiency of the waveguided photons related to re-absorption of the emitted photons by another luminescent centre. For glass-based waveguiding structures, and using a 4% figure for the front-side reflectivity, 75% for P_{TIR} corresponding to the refractive index of glass being $n = 1.5$, and by simplifying all other efficiency factors to equal 0.9, an estimate of the practical upper limit of the optical efficiency η_{opt} can be made, which is about 38%. The formula relating the parameter P_{TIR} to the refractive indices of a luminescent slab-type waveguide and its surrounding medium is available from multiple sources, e.g., [86]. Then, presuming 22% of optical-to-electric power conversion efficiency (if using monocrystalline Si cells), this upper limit for the total (device-level) LSC PCE can then be estimated to be near 8.4%. This simplified estimate of the upper limit of efficiency is unrelated to the idealised-case theory-limit calculations, but rather refers to what may be achieved in the near future, provided that luminescent materials and waveguiding structures are improving continually. This figure also relates to completely non-transparent LSC systems and will need to be reduced for all semitransparent systems, corresponding to the reduction in the incident optical power fraction being harvested. An excellent and detailed analysis of the theoretical performance limits applicable to semitransparent luminescent concentrator systems is available from [59]. In addition, PCE of near 2.5% demonstrated in a system of total (the sum of direct and diffused) visible-range transmission near 70% are shown in [55] to approach about 50% of the corresponding theory-limit system PCE, accounting for the degree of transparency and the type of solar cells used.

Equation (1) provides a useful physics-based insight into how the achievable LSC system efficiencies can continue to be improved in the future, considering that a significant scope exists for engineering new application-specific luminescent materials, optical coatings, and the internal structure of concentrator waveguides. Of special importance is the photoluminescence quantum yield of luminophores, which can potentially approach 100% in organic dyes, or even (technically) exceed 100% in "quantum cutting" inorganic phosphor types, which can emit two photons at a lower energy after single higher-energy photon excitation events. Many materials from this category can also exhibit very large Stokes shift values, thus, practically eliminating the self-absorption losses. It was predicted by the authors of [87] in 2014, that inorganic nanocrystals with high Stokes' shift and consequently low self-absorption cross section would be suitable candidates for use in highly efficient LSCs, provided that their quantum efficiency can be increased to values approaching that of dyes, which would then lead to device efficiency values beyond 10%. This prediction will likely soon be demonstrated, at least, in systems of low visible transparency, in small sample sizes, and using the record-efficiency GaAs, or multijunction record-performer cells.

Multiple developments in the materials science of advanced luminescent materials for LSC-type applications and solar window devices have been reported in the recent literature and reviews, e.g., [88–98]. The authors of [88] reviewed the current state of the field in luminescent nanomaterials for LSC, and graphically represent the photon collection mechanisms and the various loss effects as "photon destination map", which illustrates through modelling that the "escape cone losses" related to multiple total internal reflections and the associated processes, can be as high as 42% in practice, in a system where theoretically-evaluated escape-cone loss would have been near 25%. The role of reabsorption in total photon energy losses is emphasised, thus, elucidating the need to design luminescent materials with minimised self-absorption.

At present, a strong research focus and momentum are directed towards the development of advanced semiconductor nanocrystals and quantum dot (QD) materials, to enable strong reductions in self-absorption losses simultaneously with improved absorption efficiency in the near-infrared range, and to ensure greater environmental stability and light-fastness of luminophores [83,84,88–93]. Some drawbacks of inorganic QDs are discussed in [91], in particular, their susceptibility to deactivation by oxygen, often small Stokes shifts, and limited quantum efficiency, together with toxicity due to using materials, such as lead or cadmium. Organic dyes from the BASF Lumogen (Ludwigshafen, Germany) product range are still the mainstream LSC materials [91], and efforts at optimizing their concentrations and mixes are ongoing [89].

Activities aimed at achieving greater light concentration efficiency in more transparent and, importantly, larger-scale LSC devices, are also ongoing [94–97]. The authors of [95] analysed the QD-based LSCs and their maximum attainable practical concentration limits, as well as system scalability-related issues. There is currently a renewed interest in the use of upconverter-type luminophores, which were recently demonstrated to assist obtaining improved concentrator efficiencies again, achieving ~27% in PCE improvement with upconverters of only ~4% quantum yield [98].

The benefits of adding photonic mirrors to LSC structure and their effects on the optical transport phenomena within concentrators are reported in [99]. Other groups also concentrate on the development of various types of backreflector coatings, including Lambertian backreflectors [100], to improve the device PCE through enhanced partial trapping of light. Reaching the effective optical concentration ratio as high as 1.29 is reported [100]. For practically all semi-transparent solar window designs, the presence of a spectrally-selective coating backreflector is essential, since these systems are limited by their design to converting only a part of the incident solar spectrum. Since the trade-off between transparency and efficiency is fundamental, all possible approaches to maximizing the optical efficiency and the total PCE must be explored, which places a special emphasis on the structure-related photon-deflecting components (e.g., diffractive elements), working in synergy with advanced luminescent materials.

The application potential, role, and purpose of embedding the light-ray deflecting microstructures into LSC-type waveguides of solar windows are illustrated in Figure 4. Detailed analysis of the potential of diffractive and scattering optics for improving the photon collection probability at solar cells is reported in [20] and [77]. The roles of spectrally-selective backside-reflector coatings and especially the waveguiding panel thickness, in ensuring the capability of longer-range transport of the incident photons within waveguides, even in the presence of significant scattering, is also illustrated in Figure 4.

Diffractive grating-assisted ray deflection and Mie scattering processes at inorganic fluorescent pigment particles improve the probability of TIR events, thus, routing more photons towards solar cell surfaces. Highly transparent micro-scale diffraction gratings designed for use in solar windows have recently been reported [20].

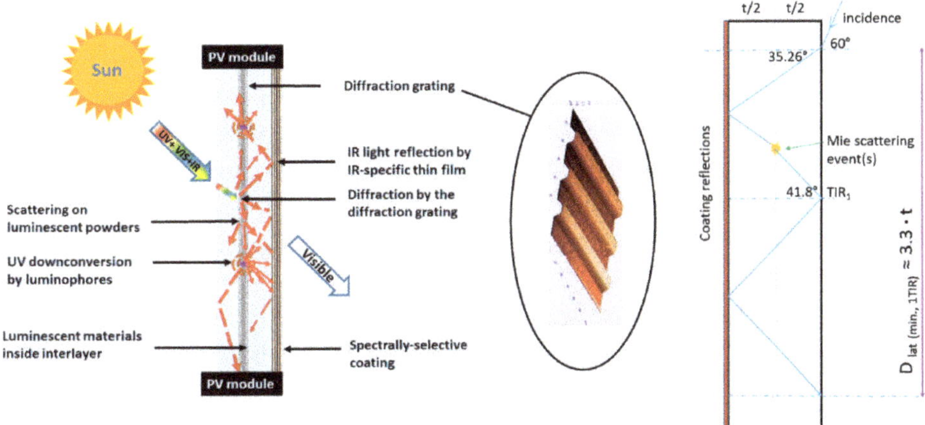

Figure 4. Semitransparent hybrid concentrator-type photovoltaic (PV) windows utilising physical mechanisms other than luminescence for inducing waveguiding-type propagation towards edge-mounted solar cells. The drawing on the right-hand side of the figure illustrates the minimum lateral displacement of photons incident at a 60° angle and travelling within a coating-assisted concentrator structure, after only a single total internal reflection (TIR) event, enabled by internal ray deflection. The minimum internal lateral displacement of partially-trapped photons incident at this angle is shown to be more than triple the system thickness dimension.

Future improvements in the photon collection efficiency and total PCE of solar windows are expected to be demonstrated in systems using optimised combinations of advanced luminescent materials and device structures. In industrialisation-ready, large-area concentrator-type systems, the designs of new glazing system structures are expected to play a leading role, even regardless of the materials selection. This is because the photon management and partial light trapping mechanisms are related fundamentally to the device geometry and the limits imposed by thermodynamics. In particular, the minimum lateral displacement of an incident light ray inside a flat slab-type concentrator system scales in proportion with the device thickness (illustrated in Figure 4). Thermodynamics-based analysis of light propagation within generic non-imaging concentrator-type systems and the photon trapping-related metrics are reported in [101,102], using Fermat's variational principle, which states that a ray of light propagates through an optical system in such a manner that the time required for it to travel from one point to another is minimised. This, together with the etendue conservation theorem [102] and the Yablonovitch limit [103], places fundamental limitations on the practically achievable light-trapping efficiencies in large-size LSC-like devices of all types. Other recently-reported important work on photon management in the presence of scattering processes [104] discloses an observation of the mean path length invariance for the photons propagating within arbitrary light-scattering media. The main finding reported was that, regardless of the location of the points of ray incidence, the system microstructure, or the angle of incidence, the mean photon propagation path-length inside the scattering medium is only governed by the system's outer boundary's geometry. This invariance imposes fundamental limitations on the internal path lengths achievable within any realistic LSC containing inorganic pigments and/or waveguide surface or volume structure imperfections. Parameters in the Equation (1), such as η_{host} and η_{TIR}, are affected substantially by these waveguide imperfections, which are present invariably in realistic devices, especially in larger-scale concentrators containing multiple interfaces, where the internal propagation path lengths start to exceed the device thickness by several times for the majority of incident photons. A particularly important result reported in [104] in relation to the theoretically-evaluated system-invariant mean value of the internal photon propagation path-length (S_{theor}) in systems containing scattering centres (in a simplified case of a single glass-air interface) is described by

$$<S_{theor}> = (4 \cdot V/\Sigma) \cdot (n_2{}^2/n_1{}^2) \tag{2}$$

where V/Σ is the system's volume-to-surface ratio, and n_1 and n_2 are the refractive indices of the outer and of the scattering regions, respectively. Therefore, in the case of rectangular slab-type waveguides representing the flat glass-based realistic inorganic pigment-containing LSC systems, this V/Σ ratio represents the system thickness, and the mean partially-trapped photon path-length is then within several times the thickness, regardless of the details of the internal microstructure. Notably, this mean internal path length invariance has been shown to apply also to the scattering process strength. Structured systems with multiple glass-air interfaces have been shown to provide mean photon path lengths somewhat in excess of that predicted by Equation (2), due to the increased probability of multiple internal reflections. Regardless of these fundamental limitations, collecting the photons efficiently at concentrator system edges, from the areas of incidence onto glass located within the range of several times the system thickness remains a valid option, which is demonstrated graphically in [77], in the presence of strong scattering at micron-scale luminophore pigment particles. The relevance of scattering as a short-range photon collection mechanism in LSC is also reported in [19,20] and discussed in [78,95].

The performance limits of luminescent solar concentrators with quantum dots in a selective-reflector-based optical cavity are evaluated in [105], where the dependency of the optical concentration on the geometric gain has also been studied. It has been reported that the best optical concentration factor performance is achievable in systems with a geometric gain of up to about 30, with further up-scaling of the device collection areas leading to diminishing returns in performance [105]. The analysis of optical losses in $CuInS_2$-based nanocrystal LSC, in terms of balancing the absorption versus scattering, is reported in [106], where the estimates for the achievable optical efficiency are provided. Overall, the findings on efficiency limits and the limiting factors in LSC performance reported in multiple recent literature sources, in conjunction with the thermodynamic limitations applicable to solar concentrators, mean that further research efforts aimed at maximizing the photon collection probability over the areas of solar window aperture closest to the solar-cell locations are necessary.

3.3. Examples of Existing and Emergent High-Transparency PV Window Technologies and Their Applications

Novel applications of transparent solar windows continue to emerge and be proposed and demonstrated. High-transparency, long product life and application-ready glass-based solar windows are still quite rare and represent the future of solar window technologies. Among these application areas so far identified are the local use of the generated and/or stored renewable energy for powering lighting systems and/or advertising display devices [55], and active in-situ control over the transparency state of "smart" glazings described in a pioneering publication [107] back in 2010. Multiple implementations of "smart windows" have been demonstrated, e.g., using polymer-dispersed liquid–crystal (PDLC) [47,108–110], or electrochromic layer-based devices [111]. As soon as the highly-transparent energy-generating solar windows begin entering widespread commercial applications in buildings, it can be predicted that their first electric loads will often be transparency-control layers embedded into the same glazing systems. This is because of both the economics of re-designing the internal electric wiring circuits in buildings required to accommodate additional loads and the ready availability of PDLC products which can be laminated into the structure of current or emergent solar window glazing designs. For these reasons, even before wiring the internal lighting circuits in buildings to be powered by highly-transparent solar windows, it can be considered logical to use the generated energy immediately at the point of generation, actively controlling the window transparency state and demonstrating the expected significant cooling-related energy savings, at least in hot climates. Figure 5 shows the smart self-powered window early-prototype demonstration experiment from October 2014, in which a PDLC layer integrated (through a glass interlayer lamination process) into a 200 mm × 200 mm LSC panel was electrically driven using the window's own electric output. The features of the electric power-generating system used in this smart window technology prototype have been disclosed in patent-related documentation [108], yet were not published academically to date.

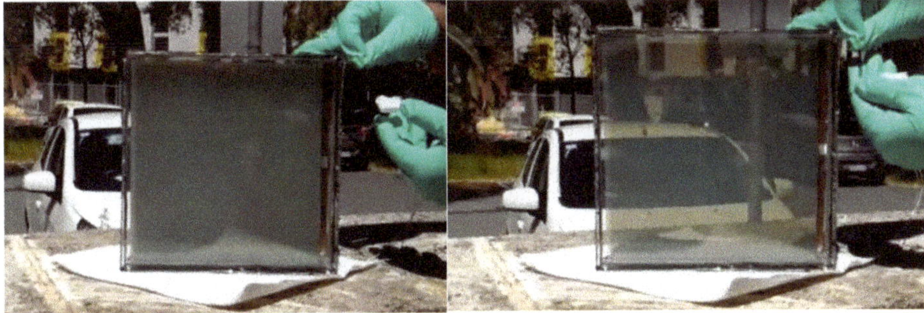

Figure 5. Transparency-controlled, self-powering PDLC-integrated "smart" solar window demonstration at Edith Cowan University (ECU) in October, 2014. A 200 mm × 200 mm hybrid concentrator-type solar window used four parallel-connected edge-mounted CuInSe$_2$ modules of size 198 mm × 25 mm, the electric output of which was sufficient to control the embedded PDLC layer transparency within a broad transparency range [108].

Significant attention from multiple research groups has been directed towards the development and demonstration of multiple window-integrated active transparency control technologies during the recent years. Among these, notable developments took place recently in the area of incorporating the photothermally switchable perovskites into window glazings [112], potentially leading to achieving substantial energy savings in buildings due to the glass transparency responding to the surface temperature changes, simultaneously with the strong energy generation potential provided by perovskite-based solar cell systems. Thermochromic and also dichroic materials-based approaches to controlling the glazing system transparency and/or apparent colour in response to illumination level changes are also gaining momentum, as well as the emergent field of smart windows overall [113–116].

It has been demonstrated ([47], and also in the Supplementary Video section of this present article) that multiple series-connected inductive loads, such as DC motor-driven ventilation fans, can be powered by a single moderate-area (500 mm × 500 mm) highly transparent solar window device, at a sufficiently fast blade rotation rate to generate audible noise heard from some distance away. Whilst the power generated in small-area clear solar windows can be very moderate (several Watts per 500 mm × 500 mm sample using edge-mounted PV cells only), there is ready availability for their practical applications and use, e.g., for charging mobile devices. Figure 6 shows a graphical summary of prototype solar window development history at ECU, Perth, Western Australia, from an early (2011) high-transparency heat mirror-coated 2 cm × 2 cm × 0.6 cm sample with stripe-shape monocrystalline cut-outs from a Si cell attached to glass edges, to more advanced and up-scaled window systems developed between 2014 and 2016. Multiple design features and characterisation results relevant to these solar window types are first reported in [20].

Figure 6. Solar window prototypes developed at Edith Cowan University (ECU between 2011–2016).

The earliest (2011) all-inorganic high-transparency (>90%) LSC prototypes integrated with Si cells (Figure 6, top left) used a single luminophore-loaded 0.5 mm interlayer connecting two 3 mm-thick quartz plates. One of these plates was coated by a spectrally-selective multilayer optical coating which reflected practically all solar near-IR light at wavelengths above 720 nm. The visible red colouration over the bottom-side solar cell surface was due to this coating reflecting a part of the visible far-red spectrum towards the cell. This small-area pre-prototype solar window sample is likely among the world's most highly-transparent energy-generating structures built using only inorganic functional materials. Later transparent solar window prototypes developed at ECU in subsequent years used various solar cell configurations and types, as well as internal micro- and macrostructures for improving the photon collection probability at solar cell surfaces, some of which are yet unpublished.

An area of organic materials-based BIPV and transparent solar windows development is undergoing rapid development currently, and new performance records are being demonstrated continually in the power conversion efficiency, thus, offering a potentially strong competition to inorganics-based concentrator-type devices and other semitransparent BIPV. Very recently, a new efficiency record was demonstrated (PCE up to 12.25%, in small-area devices), achieved by tuning the chemical structure of organic photoactive materials. This is the highest certified efficiency of organic solar cells reported to date [117,118]. However, the degree of device transparency has not been explicitly specified. It is widely expected that the optical transparency, sample size, and environmental stability properties in transparent organic cells and future BIPV systems will continue to show improvements.

In the application areas related to moving the solar window technologies from labs to markets, significant challenges still exist, regardless of the core technology platform choice, or the type of functional materials employed. This is due to the necessity of bridging multiple industry-standards-related system requirements originating from different industries. These stringent requirements range from the decades-long product lifetimes often needed, to the ability of future solar windows to resist factors, such as wind loads, water penetration, and internal condensation events. At the same time, product compliance with stringent electric wiring rules, safety-related requirements, and building construction codes are also required.

Multiple independent building-scale trials of different transparent solar window technologies are yet to be conducted to identify their application suitability in various climates and installation-area footprints, as well as the economic potential. Nevertheless, it is becoming increasingly clear, from the present-day viewpoint, that future installations of the numerous types of emergent BIPV and BAPV technologies will be numerous. This is because the benefits of the large-scale and distributed generation of energy at the locations of end use are extensive, including the capability of providing blackout resistance in city buildings and the exclusion of significant transmission-line losses. The growing field of the Internet of Things and future expected widespread application of smart interconnected (5G) devices present demanding energy use requirements, which can be addressed by widespread BIPV generation.

4. Conclusions and Outlook

The practical integration of advanced solar energy-harvesting technologies into various elements of urban landscapes, including building windows, is rapidly becoming a mainstream trend. Substantial advances have been reported in recent years both in laboratory trials, and also in commercial demonstrations of the various semi-transparent solar cell types and solar window devices. A wide range of established semi-transparent PV and BIPV technologies exists currently, providing architects and building designers with multiple choices regarding the balance between the system aesthetics, degree of transparency (or colouration type), and power generating capacity. Multiple next-generation transparent solar-cell technologies, including dye-sensitised solar cells, patterned solar panels, organic polymer-based, and perovskite-based systems remain in active stages of development and continue to demonstrate new milestones in efficiency. Highly transparent, and colour-unbiased concentrator-type solar window systems are only beginning to make their entry into industry-wide acceptance. They now provide a previously unavailable combination of up to 70% in total visible light transmission and power conversion efficiency near 2.5%, based on systems demonstrated in 2017.

Despite the fundamental trade-offs between the required control over the visual appearance, degree of transparency, and the power generating capacity intrinsic to the design of advanced BIPV, their strong potential for transforming urban landscapes and providing substantial distributed generation capacity is certain. Developments in the materials science of advanced luminophores, coupled with novel designs of LSC-type semitransparent concentrator structures add continually to the possibilities of obtaining increased power conversion efficiencies. At the same time, a substantial energy-saving potential exists, provided by solar windows, which can also control the solar heat gain in buildings and the associated thermal insulation properties. The new trends in the local utilisation of the energy generated in the distributed way by the building components include using advanced windows with active transparency control, which can contribute substantially to both personnel comfort and climate control-related energy savings. It is currently expected that multiple commercial building-based trials of the latest transparent BIPV technologies will soon be conducted, uncovering their true practical applications potential.

Supplementary Materials: The following are available online at http://www.mdpi.com/1996-1073/12/6/1080/s1, Video S1: ECU-ClearVue power generating window prototypes 2016.mp4.

Author Contributions: All authors (M.V., M.N.A. and K.A.) have contributed to the conceptualisation of this review article and data collection; M.V. analysed the data and prepared the manuscript; all authors discussed the data, graphics, and the presentation; M.N.A. contributed substantially to the data curation and the original draft preparation; M.V. and K.A. further reviewed and edited the manuscript.

Funding: This research was funded by the Australian Research Council (grants LP130100130 and LP160101589) and Edith Cowan University.

Conflicts of Interest: The authors declare no conflict of interest. The funders had no role in the design of the study; in the collection, analyses, or interpretation of data; in the writing of the manuscript, or in the decision to publish the results.

References

1. Capuano, L. *International Energy Outlook 2018 (IEO2018)*; US Energy Information Administration: Washington, DC, USA, 2018.
2. Chu, S.; Majumdar, A. Opportunities and challenges for a sustainable energy future. *Nature* **2012**, *488*, 294–303. [CrossRef] [PubMed]
3. Hoffert, M.I.; Caldeira, K.; Benford, G.; Criswell, D.R.; Green, C.; Herzog, H.; Jain, A.K.; Kheshgi, H.S.; Lackner, K.S.; Lewis, J.S.; et al. Advanced technology paths to global climate stability: Energy for a greenhouse planet. *Science* **2002**, *298*, 981–987. [CrossRef]
4. Poizot, P.; Dolhem, F. Clean energy new deal for a sustainable world: From non-CO_2 generating energy sources to greener electrochemical storage devices. *Energy Environ. Sci.* **2011**, *4*, 2003–2019. [CrossRef]

5. Cao, X.; Xilei, D.; Liu, J. Building energy-consumption status worldwide and the state-of-the-art technologies for zero-energy buildings during the past decade. *Energy Build.* **2016**, *128*, 198–213. [CrossRef]
6. Schellnhuber, H.J.; van der Hoeven, M.; Bastioli, C.; Ekins, P.; Jaczewska, B.; Kux, B.; Thimann, C.; Tubiana, L.; Wanngård, K.; Slob, A.; et al. *Final Report of the High-Level Panel of the European Decarbonisation Pathways Initiative*; European Commission: Brussels, Belgium, 2018; ISBN 978-92-79-96827-3.
7. Long, L.; Ye, H. How to be smart and energy efficient: A general discussion on thermochromic windows. *Sci. Rep.* **2014**, *4*, 6427. [CrossRef] [PubMed]
8. Gao, Y.; Luo, H.; Zhang, Z.; Kang, L.; Chen, Z.; Du, J.; Kanehira, M.; Cao, C. Nanoceramic VO_2 thermochromic smart glass: A review on progress in solution processing. *Nano Energy* **2012**, *1*, 221–246. [CrossRef]
9. Llordes, A.; Garcia, G.; Gazquez, J.; Milliron, D.J. Tunable near-infrared and visible-light transmittance in nanocrystal-in-glass composites. *Nature* **2013**, *500*, 323–326. [CrossRef]
10. Raman, A.P.; Anoma, M.A.; Zhu, L.; Rephaeli, E.; Fan, S. Passive radiative cooling below ambient air temperature under direct sunlight. *Nature* **2014**, *515*, 540–544. [CrossRef]
11. Fu, Y.; Yang, J.; Su, Y.S.; Du, W.; Ma, Y.G. Daytime passive radiative cooler using porous alumina. *Sol. Energy Mater. Sol. Cells* **2019**, *191*, 50–54. [CrossRef]
12. Wang, J.; Zhang, L.; Yu, L.; Jiao, Z.; Xie, H.; Lou, X.W.; Sun, X.W. A bi-functional device for self-powered electrochromic window and self-rechargeable transparent battery applications. *Nat. Commun.* **2014**, *5*, 4921. [CrossRef]
13. Wang, S.; Liu, M.; Kong, L.; Long, Y.; Jiang, X.; Yu, A. Recent progress in VO_2 smart coatings: Strategies to improve the thermochromic properties. *Prog. Mater. Sci.* **2016**, *81*, 1–54. [CrossRef]
14. Torcellini, P.; Pless, S.; Deru, M.; Crawley, D. *Zero Energy Buildings: A Critical Look at the Definition*; National Renewable Energy Laboratory and Department of Energy: Golden, CO, USA, 2006.
15. Groth, R.; Reichelt, W. Gold coated glass in the building industry. *Gold Bull.* **1974**, *7*, 62–68. [CrossRef]
16. Lampert, C.M. Heat mirror coatings for energy conserving windows. *Sol. Energy Mater.* **1981**, *6*, 1–41. [CrossRef]
17. Dalapati, G.K.; Kushwaha, A.K.; Sharma, M.; Suresh, V.; Shannigrahi, S.; Zhuk, S.; Masudy-Panah, S. Transparent heat regulating (THR) materials and coatings for energy saving window applications: Impact of materials design, micro-structural, and interface quality on the THR performance. *Prog. Mater. Sci.* **2018**, *95*, 42–131. [CrossRef]
18. Rezaei, S.D.; Shannigrahi, S.; Ramakrishna, S. A review of conventional, advanced, and smart glazing technologies and materials for improving indoor environment. *Sol. Energy Mater. Sol. Cells* **2017**, *159*, 26–51. [CrossRef]
19. Alghamedi, R.; Vasiliev, M.; Nur-E-Alam, M.; Alameh, K. Spectrally-Selective All-Inorganic Scattering Luminophores For Solar Energy-Harvesting Clear Glass Windows. *Sci. Rep.* **2014**, *4*, 6632. [CrossRef]
20. Vasiliev, M.; Alghamedi, R.; Nur-E-Alam, M.; Alameh, K. Photonic microstructures for energy-generating clear glass and net-zero energy buildings. *Sci. Rep.* **2016**, *6*, 31831. [CrossRef]
21. Rosenberg, V.; Vasiliev, M.; Alameh, K. A Spectrally Selective Luminescence Concentrator Panel with a Photovoltaic Cell. EP Patent 2 726 920 B1, 23 August 2017.
22. Jelle, B.P.; Breivik, C. The path to the building integrated photovoltaics of tomorrow. *Energy Procedia* **2012**, *20*, 78–87. [CrossRef]
23. Debbarma, M.; Sudhakar, K.; Baredar, P. Comparison of BIPV and BIPVT: A review. *Resour.-Effic. Technol.* **2017**, *3*, 263–271. [CrossRef]
24. Biyik, E.; Araz, M.; Hepbasli, A.; Shahrestani, M.; Yao, R.; Shao, L.; Essah, E.; Oliveira, A.C.; del Caño, T.; Rico, E.; et al. A key review of building integrated photovoltaic (BIPV) systems. *Eng. Sci. Technol. Int. J.* **2017**, *20*, 833–858. [CrossRef]
25. Tripathy, M.; Sadhu, P.K.; Panda, S.K. A critical review on building integrated photovoltaic products and their applications. *Renew. Sustain. Energy Rev.* **2016**, *61*, 451–465. [CrossRef]
26. Extance, A. The Dawn of Solar Windows, IEEE Spectrum 2018. Online Publication. Available online: https://spectrum.ieee.org/energy/renewables/the-dawn-of-solar-windows (accessed on 10 February 2018).
27. International Energy Agency—Photovoltaic Power Systems Programme Annual Report 2016. Available online: https://docplayer.net/52701446-P-h-o-t-o-v-o-l-t-a-i-c-p-o-w-e-r-s-y-s-t-e-m-s-p-r-o-g-r-a-m-m-e-annual-report.html (accessed on 26 November 2018).

28. Building Integrated Photovoltaics: Product Overview for Solar Building Skins, Status Report 2017. Available online: https://www.seac.cc/wp-content/uploads/2017/11/171102_SUPSI_BIPV.pdf (accessed on 26 November 2018).

29. Attoye, D.E.; Aoul, K.A.T.; Hassan, A. A Review on Building Integrated Photovoltaic Façade Customization Potentials. *Sustainability* **2017**, *9*, 2287. [CrossRef]

30. Cornaro, C.; Basciano, G.; Puggioni, V.A.; Pierro, M. Energy Saving Assessment of Semi-Transparent Photovoltaic Modules Integrated into NZEB. *Buildings* **2017**, *7*, 9. [CrossRef]

31. Shin, D.H.; Choi, S.-H. Recent Studies of Semitransparent Solar Cells. *Coatings* **2018**, *8*, 329. [CrossRef]

32. Lee, K.-T.; Guo, L.J.; Park, H.J. Neutral- and Multi-Colored Semitransparent Perovskite Solar Cells. *Molecules* **2016**, *21*, 475. [CrossRef] [PubMed]

33. The School with the Largest Solar Facade in the World. Available online: https://phys.org/news/2017-02-school-largest-solar-facade-world.html (accessed on 26 November 2018).

34. Massive Solar Façade for Swiss Convention Centre. Available online: https://www.forumforthefuture.org/greenfutures/articles/massive-solar-façade-swiss-convention-centre (accessed on 2 May 2018).

35. Barman, S.; Chowdhury, A.; Mathur, S.; Mathur, J. Assessment of the efficiency of window integrated CdTe based semitransparent photovoltaic module. *Sustain. Cities Soc.* **2018**, *37*, 250–262. [CrossRef]

36. Weber, W.H.; Lambe, J. Luminescent greenhouse collector for solar radiation. *Appl. Opt.* **1976**, *15*, 2299–2300. [CrossRef]

37. Goetzberger, A.; Greube, W. Solar-energy conversion with fluorescent collectors. *Appl. Phys.* **1977**, *14*, 123–139. [CrossRef]

38. Debije, M.G.; Verbunt, P.P.C. Thirty Years of Luminescent Solar Concentrator Research: Solar Energy for the Built Environment. *Adv. Energy Mater.* **2012**, *2*, 12–35. [CrossRef]

39. Mazzaro, R.; Vomiero, A. The Renaissance of Luminescent Solar Concentrators: The Role of Inorganic Nanomaterials. *Adv. Energy Mater.* **2018**, *8*, 1801903. [CrossRef]

40. Reinders, A.; Kishore, R.; Slooff, L.; Eggink, W. Luminescent solar concentrator photovoltaic designs. *Jpn. J. Appl. Phys.* **2018**, *57*, 08RD10. [CrossRef]

41. Meinardi, F.; Colombo, A.; Velizhanin, K.A.; Simonutti, R.; Lorenzon, M.; Beverina, L.; Viswanatha, R.; Klimov, V.I.; Brovelli, S. Large-area luminescent solar concentrators based on 'Stokes-shift-engineered' nanocrystals in a mass-polymerized PMMA matrix. *Nat. Photonics* **2014**, *8*, 392–399. [CrossRef]

42. Bergren, M.R.; Makarov, N.S.; Ramasamy, K.; Jackson, A.; Guglielmetti, R.; McDaniel, H. High-Performance CuInS$_2$ Quantum Dot Laminated Glass Luminescent Solar Concentrators for Windows. *ACS Energy Lett.* **2018**, *3*, 520–525. [CrossRef]

43. Zhao, Y.; Meek, G.A.; Levine, B.G.; Lunt, R.R. Near-Infrared Harvesting Transparent Luminescent Solar Concentrators. *Adv. Opt. Mater.* **2014**, *2*, 606–611. [CrossRef]

44. Traverse, C.J.; Pandey, R.; Barr, M.C.; Lunt, R.R. Emergence of highly transparent photovoltaics for distributed applications. *Nat. Energy* **2017**, *2*, 849–860. [CrossRef]

45. So, F. Organics: Materials and Energy. In *World Scientific Handbook of Organic Optoelectronic Devices*; Transparent Photovoltaics; World Scientific: Singapore, 2018; Chapter 11; pp. 445–499.

46. Husain, A.A.F.; Hasan, W.Z.W.; Shafie, S.; Hamidon, M.N.; Pandey, S.S. A review of transparent solar photovoltaic technologies. *Renew. Sustain. Energy Rev.* **2018**, *94*, 779–791. [CrossRef]

47. Dalapati, G.K.; Sharma, M. *Heat Regulating Materials for Energy Savings and Generation: Materials Design, Process and Implementation*; CWP Publishing: Orange, Australia, 2019. Available online: https://centralwestpublishing.com/specialty-materials/ (accessed on 19 March 2019).

48. Gates, B. Energy Innovation—Why We Need It and How to Get It. Published by Breakthrough Energy Coalition. 2015. Available online: https://www.google.com/url?sa=t&rct=j&q=&esrc=s&source=web&cd=1&cad=rja&uact=8&ved=2ahUKEwjQjvmfiaPeAhVWdCsKHVsMCbQQFjAAegQICRAC&url=https%3A%2F%2Fwww.gatesnotes.com%2F~{}%2Fmedia%2FFiles%2FEnergy%2FEnergy_Innovation_Nov_30_2015.pdf%3Fla%3Den&usg=AOvVaw1VZwJaefqQT1p4Qrm-UzYm (accessed on 19 March 2019).

49. Ryan, D. Solar Panels Replaced Tarmac on a Motorway—Here Are the Results. 2018. Available online: https://phys.org/news/2018-09-solar-panels-tarmac-motorwayhere-results.html (accessed on 6 December 2018).

50. Avancis PowerMax Skala Datasheet. Available online: https://www.avancis.de/en/products/powermaxr-skala-40/ (accessed on 5 December 2018).

51. Sunjoule Product Brochure by Asahi Glass Corp. p. 11. Available online: http://www.agc-solar.com/agc-solar-products/bipv.html (accessed on 5 December 2018).

52. Onyx Solar Amorphous Silicon PV Glass Specifications—High Transparency. Available online: https://www.onyxsolar.com/product-services/technical-specifications (accessed on 5 December 2018).

53. Hanergy Product Manual 141129 Section 1.3.1 (2018). p. 11. Available online: https://www.slideshare.net/RonaldKranenberg/hanergyproductbrochure141129 (accessed on 5 December 2018).

54. Solaronix Solar Cells Brochure. Available online: http://solaronix.com/solarcells/ (accessed on 11 December 2018).

55. Vasiliev, M.; Alameh, K.; Nur-E-Alam, M. Spectrally-Selective Energy-Harvesting Solar Windows for Public Infrastructure Applications. *Appl. Sci.* **2018**, *8*, 849. [CrossRef]

56. CdTe-Based Semitransparent BIPV Products from Xiamen Solar First Energy Technology Co., Ltd. Available online: https://www.alibaba.com/product-detail/High-Transparent-Asi-frameless-CdTe-48W_60541808069.html (accessed on 11 December 2018).

57. Slooff, L.H.; Bende, E.E.; Burgers, A.R.; Budel, T.; Pravettoni, M.; Kenny, R.P.; Dunlop, E.D.; Büchtemann, A. A luminescent solar concentrator with 7.1% power conversion efficiency. *Phys. Status Solidi (RRL)* **2008**, *2*, 257–259. [CrossRef]

58. Tummeltshammer, C.; Taylor, A.; Kenyon, A.J.; Papakonstantinou, I. Losses in luminescent solar concentrators unveiled. *Sol. Energy Mater. Sol. Cells* **2016**, *144*, 40–47. [CrossRef]

59. Yang, C.; Lunt, R.R. Limits of visibly transparent luminescent solar concentrators. *Adv. Opt. Mater.* **2017**, *5*, 1600851. [CrossRef]

60. Wang, W.; Yan, C.; Lau, T.-K.; Wang, J.; Liu, K.; Fan, Y.; Lu, X.; Zhan, X. Fused Hexacyclic Nonfullerene Acceptor with Strong Near-Infrared Absorption for Semitransparent Organic Solar Cells with 9.77% Efficiency. *Adv. Mater.* **2017**, *29*, 1701308. [CrossRef] [PubMed]

61. Chen, C.-C.; Dou, L.; Zhu, R.; Chung, C.-H.; Song, T.-B.; Zheng, Y.B.; Hawks, S.; Li, G.; Weiss, P.S.; Yang, Y. Visibly Transparent Polymer Solar Cells Produced by Solution Processing. *ACS Nano* **2012**, *6*, 7185–7190. [CrossRef]

62. VELUX Solar Powered Skylight (VSS). Available online: https://www.velux.com.au/products/skylights/vss%20solar%20skylight (accessed on 19 March 2019).

63. Physee PowerWindow Products. Available online: https://www.physee.eu/products (accessed on 11 December 2018).

64. GlassToPower Technology Description. Available online: http://www.glasstopower.com/la-tecnologia-en/ (accessed on 12 December 2018).

65. Ito, S.; Chen, P.; Comte, P.; Nazeeruddin, M.K.; Liska, P.; Péchy, P.; Grätzel, M. Fabrication of Screen-Printing Pastes from TiO$_2$ Powders for Dye-Sensitised Solar Cells. *Prog. Photovolt. Res. Appl.* **2007**, *15*, 603–612. [CrossRef]

66. Lunt, R.R.; Bulovic, V. Transparent, near-infrared organic photovoltaic solar cells for window and energy-scavenging applications. *Appl. Phys. Lett.* **2011**, *98*, 113305. [CrossRef]

67. Bailie, C.D.; Christoforo, M.G.; Mailoa, J.P.; Bowring, A.R.; Unger, E.L.; Nguyen, W.H.; Burschka, J.; Pellet, N.; Lee, J.Z.; Grätzel, M.; et al. Semi-transparent perovskite solar cells for tandems with silicon and CIGS. *Energy Environ. Sci.* **2015**, *8*, 956–963. [CrossRef]

68. Roldán-Carmona, C.; Malinkiewicz, O.; Betancur, R.; Longo, G.; Momblona, C.; Jaramillo, F.; Camacho, L.; Bolink, H.J. High efficiency single-junction semitransparent perovskite solar cells. *Energy Environ. Sci.* **2014**, *7*, 2968–2973. [CrossRef]

69. Zhang, X.; Hägglund, C.; Johansson, M.B.; Sveinbjörnsson, K.; Johansson, E.M.J. Fine Tuned Nanolayered Metal/Metal Oxide Electrode for Semitransparent Colloidal Quantum Dot Solar Cells. *Adv. Funct. Mater.* **2016**, *26*, 1921–1929. [CrossRef]

70. Polysolar BIPV Solar Glass Specifications. Available online: http://www.polysolar.co.uk/Technology/PS-CT (accessed on 14 December 2018).

71. Qiu, L.; Liu, Z.; Ono, L.K.; Jiang, Y.; Son, D.-Y.; Hawash, Z.; He, S.; Qi, Y. Scalable Fabrication of Stable High Efficiency Perovskite Solar Cells and Modules Utilizing Room Temperature Sputtered SnO$_2$ Electron Transport Layer. *Adv. Funct. Mater.* **2018**, 1806779. [CrossRef]

72. Oxford PV Sets World Record for Perovskite Solar Cell. Available online: https://www.oxfordpv.com/news/oxford-pv-sets-world-record-perovskite-solar-cell (accessed on 24 January 2019).

73. Researchers Show World-Record 20.9% Perovskite Cell, New SSG Process Increasing Stability. Available online: https://www.pv-magazine.com/2018/07/05/researchers-show-world-record-20-9-perovskite-cell-new-ssg-process-increasing-stability/ (accessed on 24 January 2019).

74. Wang, S.; Jiang, Y.; Juarez-Perez, E.-J.; Ono, L.K.; Qi, Y. Accelerated degradation of methylammonium lead iodide perovskites induced by exposure to iodine vapour. *Nat. Energy* **2016**, *2*, 16195. [CrossRef]

75. Soleimanioun, N.; Rani, M.; Sharma, S.; Kumar, A.; Tripathi, S.K. Binary metal zinc-lead perovskite built-in air ambient: Towards lead-less and stable perovskite materials. *Sol. Energy Mater. Sol. Cells* **2019**, *191*, 339–344. [CrossRef]

76. Yan, X.; Hu, S.; Zhang, Y.; Li, H.; Sheng, C. Methylammonium acetate as an additive to improve performance and eliminate J-V hysteresis in 2D homologous organic-inorganic perovskite solar cells. *Sol. Energy Mater. Sol. Cells* **2019**, *191*, 283–289. [CrossRef]

77. Vasiliev, M.; Alameh, K.; Badshah, M.A.; Kim, S.-M.; Nur-E-Alam, M. Semi-Transparent Energy-Harvesting Solar Concentrator Windows Employing Infrared Transmission-Enhanced Glass and Large-Area Microstructured Diffractive Elements. *Photonics* **2018**, *5*, 25. [CrossRef]

78. Liu, H.; Li, S.; Chen, W.; Wang, D.; Li, C.; Wu, D.; Hao, J.; Zhou, Z.; Wang, X.; Wang, K. Scattering enhanced quantum dots based luminescent solar concentrators by silica microparticles. *Sol. Energy Mater. Sol. Cells* **2018**, *179*, 380–385. [CrossRef]

79. Leow, S.W.; Corrado, C.; Osborn, M.; Isaacson, M.; Alers, G.; Carter, S.A. Analyzing luminescent solar concentrators with front-facing photovoltaic cells using weighted Monte Carlo ray tracing. *J. Appl. Phys.* **2013**, *113*, 214510. [CrossRef]

80. Corrado, C.; Leow, S.W.; Osborn, M.; Carbone, I.; Hellier, K.; Short, M.; Alers, G.; Carter, S.A. Power generation study of luminescent solar concentrator greenhouse. *J. Renew. Sustain. Energy* **2016**, *8*, 043502. [CrossRef]

81. Reinders, A.; Debije, M.G.; Rosemann, A. Measured Efficiency of a Luminescent Solar Concentrator PV Module Called Leaf Roof. *IEEE J. Photovolt.* **2017**, *7*, 1663–1666. [CrossRef]

82. Desmet, L.; Ras, A.J.M.; de Boer, D.K.G.; Debije, M.G. Monocrystalline silicon photovoltaic luminescent solar concentrator with 4.2% power conversion efficiency. *Opt. Lett.* **2012**, *37*, 3087–3089. [CrossRef] [PubMed]

83. Wu, K.; Li, H.; Klimov, V.I. Tandem luminescent solar concentrators based on engineered quantum dots. *Nat. Photonics* **2018**, *12*, 105–110. [CrossRef]

84. Meinardi, F.; Ehrenberg, S.; Dhamo, L.; Carulli, F.; Mauri, M.; Bruni, F.; Simonutti, R.; Kortshagen, U.; Brovelli, S. Highly efficient luminescent solar concentrators based on earth-abundant indirect-bandgap silicon quantum dots. *Nat. Photonics* **2017**, *11*, 177–185. [CrossRef]

85. Aste, N.; Tagliabue, L.C.; Del Pero, C.; Testa, D.; Fusco, R. Performance analysis of a large-area luminescent solar concentrator module. *Renew. Energy* **2015**, *76*, 330–337. [CrossRef]

86. Xu, L.; Yao, Y.; Bronstein, N.D.; Li, L.; Alivisatos, A.P.; Nuzzo, R.G. Enhanced Photon Collection in Luminescent Solar Concentrators with Distributed Bragg Reflectors. *ACS Photonics* **2016**, *3*, 278–285. [CrossRef]

87. Van Sark, W.G.J.H.M. Will luminescent solar concentrators surpass the 10% device efficiency limit? *SPIE Newsroom* **2014**, *9178*. [CrossRef]

88. Moraitisa, P.; Schroppb, R.E.I.; van Sark, W.G.J.H.M. Nanoparticles for Luminescent Solar Concentrators—A review. *Opt. Mater.* **2018**, *84*, 636–645. [CrossRef]

89. Krumera, Z.; van Sark, W.G.J.H.M.; Schroppc, R.E.I.; de Mello Donegá, C. Compensation of self-absorption losses in luminescent solar concentrators by increasing luminophore concentration. *Sol. Energy Mater. Sol. Cells* **2017**, *167*, 133–139. [CrossRef]

90. Zhao, Y.; Lunt, R.R. Transparent luminescent solar concentrators for large-area solar windows enabled by massive stokes-shift nanocluster phosphors. *Adv. Energy Mater.* **2013**, *3*, 1143–1148. [CrossRef]

91. Debije, M. Semiconductor solution. *Nat. Photonics* **2017**, *11*, 143–144. [CrossRef]

92. Ten Kate, O.M.; Krämer, K.W.; Van der Kolk, E. Efficient luminescent solar concentrators based on self-absorption free, Tm^{2+} doped halides. *Sol. Energy Mater. Sol. Cells* **2015**, *140*, 115–120. [CrossRef]

93. Erickson, C.S.; Bradshaw, L.R.; McDowall, S.; Gilbertson, J.D.; Gamelin, D.R.; Patrick, D.L. Zero-Reabsorption Doped-Nanocrystal Luminescent Solar Concentrators. *ACS Nano* **2014**, *8*, 3461–3467. [CrossRef]

94. Li, H.; Wu, K.; Lim, J.; Song, H.-J.; Klimov, V.I. Doctor-blade deposition of quantum dots onto standard window glass for low-loss large-area luminescent solar concentrators. *Nat. Energy* **2016**, *1*, 16157. [CrossRef]

95. Klimov, V.I.; Baker, T.A.; Lim, J.; Velizhanin, K.A.; McDaniel, H. Quality Factor of Luminescent Solar Concentrators and Practical Concentration Limits Attainable with Semiconductor Quantum Dots. *ACS Photonics* **2016**, *3*, 1138–1148. [CrossRef]

96. Zhao, H.; Zhou, Y.; Benettia, D.; Ma, D.; Rosei, F. Perovskite quantum dots integrated in large-area luminescent solar concentrators. *Nano Energy* **2017**, *37*, 214–223. [CrossRef]

97. Gong, X.; Ma, W.; Li, Y.; Zhong, L.; Li, W.; Zhao, X. Fabrication of high-performance luminescent solar concentrators using N-doped carbon dots/PMMA mixed matrix slab. *Org. Electron.* **2018**, *63*, 237–243. [CrossRef]

98. Ha, S.-J.; Kang, J.-H.; Choi, D.H.; Nam, S.K.; Reichmanis, E.; Moon, J.H. Upconversion-Assisted Dual-Band Luminescent Solar Concentrator Coupled for High Power Conversion Efficiency Photovoltaic Systems. *ACS Photonics* **2018**, *5*, 3621–3627. [CrossRef]

99. Connell, R.; Ferry, V.E. Integrating Photonics with Luminescent Solar Concentrators: Optical Transport in the Presence of Photonic Mirrors. *J. Phys. Chem. C* **2016**, *120*, 20991–20997. [CrossRef]

100. Liu, X.; Wu, Y.; Hou, X.; Liu, H. Investigation of the Optical Performance of a Novel Planar Static PV Concentrator with Lambertian Rear Reflectors. *Buildings* **2017**, *7*, 88. [CrossRef]

101. Rau, U.; Paetzold, U.W.; Kirchartz, T. Thermodynamics of light management in photovoltaic devices. *Phys. Rev. B* **2014**, *90*, 035211. [CrossRef]

102. Shatz, N.; Bortz, J.; Winston, R. Thermodynamic efficiency of solar concentrators. *Opt. Express* **2010**, *18*, A5–A16. [CrossRef]

103. Yablonovitch, E. Thermodynamics of the fluorescent planar concentrator. *J. Opt. Soc. Am.* **1980**, *70*, 1362–1363. [CrossRef]

104. Savo, R.; Pierrat, R.; Najar, U.; Carminati, R.; Rotter, S.; Gigan, S. Observation of mean path length invariance in light-scattering media. *Science* **2017**, *358*, 765–768. [CrossRef]

105. Song, H.-J.; Jeong, B.G.; Lim, J.; Lee, D.C.; Wan Ki Bae, W.K.; Klimov, V.I. Performance Limits of Luminescent Solar Concentrators Tested with Seed/Quantum-Well Quantum Dots in a Selective-Reflector-Based Optical Cavity. *Nano Lett.* **2018**, *18*, 395–404. [CrossRef] [PubMed]

106. Sumner, R.; Eiselt, S.; Kilburn, T.B.; Erickson, C.; Carlson, B.; Gamelin, D.R.; McDowall, S.; Patrick, D.L. Analysis of Optical Losses in High-Efficiency CuInS$_2$-Based Nanocrystal Luminescent Solar Concentrators: Balancing Absorption versus Scattering. *J. Phys. Chem. C* **2017**, *121*, 3252–3260. [CrossRef]

107. Debije, M.G. Solar Energy Collectors with Tunable Transmission. *Adv. Funct. Mater.* **2010**, *20*, 1498–1502. [CrossRef]

108. Vasiliev, M.; Alameh, K.; Rosenberg, V. A Device for Generating Electric Energy. U.S. Patent 20160204297A1, 14 July 2016.

109. Mateen, F.; Oh, H.; Jung, W.; Lee, S.Y.; Kikuchi, H.; Hong, S.-K. Polymer dispersed liquid crystal device with integrated luminescent solar concentrator. *Liq. Cryst.* **2018**, *45*, 498–506. [CrossRef]

110. Sol, J.A.H.P.; Timmermans, G.H.; van Breugel, A.J.; Schenning, A.P.H.J.; Debije, M.G. Multistate Luminescent Solar Concentrator "Smart" Windows. *Adv. Energy Mater.* **2018**, *8*, 1702922. [CrossRef]

111. Fernandes, M.; Freitas, V.; Pereira, S.; Leones, R.; Silva, M.M.; Carlos, L.D.; Fortunato, E.; Rute, A.S.; Ferreira, R.A.S.; Rego, R.; et al. Luminescent Electrochromic Devices for Smart Windows of Energy-Efficient Buildings. *Energies* **2018**, *11*, 3513. [CrossRef]

112. Wheeler, L.M.; Moore, D.T.; Ihly, R.; Stanton, N.J.; Miller, E.M.; Tenent, R.C.; Blackburn, J.L.; Neale, N.R. Switchable photovoltaic windows enabled by reversible photothermal complex dissociation from methylammonium lead iodide. *Nat. Commun.* **2017**, *8*, 1722. [CrossRef] [PubMed]

113. Zhou, J.; Gao, Y.; Zhang, Z.; Luo, H.; Cao, C.; Chen, Z.; Dai, L.; Liu, X. VO$_2$ thermochromic smart window for energy savings and generation. *Sci. Rep.* **2013**, *3*, 3029. [CrossRef] [PubMed]

114. Garshasbi, S.; Santamouris, M. Using advanced thermochromic technologies in the built environment: Recent development and potential to decrease the energy consumption and fight urban overheating. *Sol. Energy Mater. Sol. Cells* **2019**, *191*, 21–32. [CrossRef]

115. Casini, M. Active dynamic windows for buildings: A review. *Renew. Energy* **2018**, *119*, 923–934. [CrossRef]

116. Yoo, G.Y.; Jeong, J.-S.; Lee, S.; Lee, Y.; Yoon, H.C.; Van Ben Chu, V.B.; Park, G.S.; Hwang, Y.J.; Kim, W.; Min, B.K.; et al. Multiple-Color-Generating Cu(In,Ga)(S,Se)$_2$ Thin-Film Solar Cells via Dichroic Film Incorporation for Power-Generating Window Applications. *ACS Appl. Mater. Interfaces* **2017**, *9*, 14817–14826. [CrossRef]

117. Fan, B.; Du, X.; Liu, F.; Zhong, W.; Ying, L.; Xie, R.; Tang, X.; An, K.; Xin, J.; Li, N.; et al. Fine-tuning of the chemical structure of photoactive materials for highly efficient organic photovoltaics. *Nat. Energy* **2018**, *3*, 1051–1058. [CrossRef]
118. Researchers Achieve Highest Certified Efficiency of Organic Solar Cells to Date. 2018. Available online: https://phys.org/news/2018-11-fine-tuning-record-breaking.html (accessed on 10 December 2018).

Review

Review on Building-Integrated Photovoltaics Electrical System Requirements and Module-Integrated Converter Recommendations

Simon Ravyts [1,2,*], Mauricio Dalla Vecchia [1,2], Giel Van den Broeck [1,2] and Johan Driesen [1,2]

[1] Department of Electrical Engineering, ESAT, KU Leuven, Kasteelpark Arenberg 10, 3001 Leuven, Belgium; mauricio.dallavecchia@kuleuven.be (M.D.V.); giel.vandenbroeck@kuleuven.be (G.V.d.B.); johan.driesen@kuleuven.be (J.D.)

[2] EnergyVille, Thor Park 8310, 3600 Genk, Belgium

* Correspondence: simon.ravyts@kuleuven.be

Received: 26 March 2019 ; Accepted: 17 April 2019 ; Published: 23 April 2019

Abstract: Since building-integrated photovoltaic (BIPV) modules are typically installed during, not after, the construction phase, BIPVs have a profound impact compared to conventional building-applied photovoltaics on the electrical installation and construction planning of a building. As the cost of BIPV modules decreases over time, the impact of electrical system architecture and converters will become more prevalent in the overall cost of the system. This manuscript provides an overview of potential BIPV electrical architectures. System-level criteria for BIPV installations are established, thus providing a reference framework to compare electrical architectures. To achieve modularity and to minimize engineering costs, module-level DC/DC converters preinstalled in the BIPV module turned out to be the best solution. The second part of this paper establishes converter-level requirements, derived and related to the BIPV system. These include measures to increase the converter fault tolerance for extended availability and to ensure essential safety features.

Keywords: PV; BIPV; LVDC; DC/DC module-level converters

---

1. Introduction

1.1. Motivation

Building-integrated photovoltaic (BIPV) systems consist of solar photovoltaic (PV) cells and modules that are integrated in the building envelope as part of the building structure, replacing conventional building materials [1,2]. BIPV is now being proposed as an economically viable solution to the increasing demand for renewable electricity generation, since the relatively minor added cost of PV cells to the overall building component's cost results in conceivable payback times [3,4]. Furthermore, developments in thin-film PV technology reduce the costs of adding PV to structural elements even further [5,6].

The use of BIPV is encouraged by the European Strategic Energy Technology (SET) Plan [7] and the European Energy Performance of Buildings Directive (EPBD) [8]. The EPBD requires that all new buildings in the 28 member states are near Zero Energy Buildings (NZEB) from 2020 on. The implementation can be a combined result of reducing the energy demand and increasing the energy generation on site. In high-rise buildings, the amount of roof surface where PV panels can be placed might be unsufficient to cover the demand of the building. Placing PV in the façades offers a solution to this [9,10]. A high amount of research is conducted towards the different aspects of BIPV, such as BIPV Thermal (BIPVT) installations [11–13], a life-cycle analysis of BIPV installations [14,15], ab optimal design to match the electric loads in NZEB [16], the refurbishment and renovation of older buildings using BIPV [17–19], the thermal impact on the building [20–23], novel PV materials for use in BIPV products [24–27], the role of Building Information Management (BIM) in the design of new buildings with BIPV [28–30], and specific case studies [31–35]. This manuscript focuses on the electrical installation aspects of façade BIPV modules where a high degree of modularity is envisioned. At first sight, the electrical installation of BIPV systems does not seem to differ from building-applied photovoltaics (BAPV). This paper will highlight that the expectations and boundary conditions are different and lead to specific requirements for the design of new power electronics converters.

A wide variety of BIPV modules is commercially available on the market [3,36–39]. In general, two classes of BIPV modules are distinguished, namely roofing BIPV modules [40] and façade BIPV modules [3,41]. The first category consists of PV modules that are part of the roof structure of a building, comprising in-roof systems, full roof solutions, and solar tiles. The second category further distinguishes cold and warm façades, depending on whether the BIPV modules contain a ventilated air gap or not. Additionally, accessories exist such as parapets, balconies, and solar shadings not belonging directly to the building skin [42].

More than BAPV systems, BIPV systems have a profound impact on the electrical installation of the building and on the planning of the construction works, as the BIPV modules are installed during the construction phase and not fitted after construction has completed. Hence, the way in which the BIPV modules are interconnected and converter-interfaced is important to consider in order to minimize the additional required installation time and system engineering effort. In that regard, module-level converters (MLCs) that are factory preinstalled in the BIPV modules are promising [43]. Their main advantage of modularity prevails over the use of conventional string-level inverters, which are usually praised for their lower system cost [44]. Furthermore, adopting a DC instead of an AC distribution architecture simplifies the design of the module-level converters, resulting in an increased compactness, efficiency, and reliability [45,46]. Apart from that, as the cost of BIPV modules reduces further over time, the impact of the electrical installation architecture and the converters becomes more prevalent in the overall cost of the system.

Besides modularity, module-level converters reduce the impact of partial shading and are capable of supporting a wide variety of BIPV modules and associated electrical specifications [36,47–49]. Especially partial shading, leading to different I-V characteristics and maximum power points (MPP) of neighbouring modules is more prevalent in BIPV systems as compared to BAPV [50–53].

However, preinstalling module-level converters in BIPV modules leads to higher required levels of fault tolerance of the converters as it is undesirable or practically infeasible to replace the converter after an internal failure. Therefore, this paper discusses the possible failure modes of module-level converters, including its causes, consequences, and detection methods, and introduces techniques to ensure the fail-safe operation of the module-level converter if necessary. Additionally, the need for non-isolated or isolated module-level converters will be addressed.

1.2. Research Questions and Objectives

The research questions in this paper are (1) Given the difference with BAPV, what are the specific requirements of a BIPV electrical installation? General ideas will be translated to concrete evaluation points that serve as Key Performance Indicators (KPIs); (2) What are the advantages of Low-Voltage DC (LVDC) grids compared to traditional AC grids to electrically interconnect BIPV modules?; (3) Can the electrical requirements be translated to practical converter design recommendations?; and (4) What is the impact of the LVDC grid configuration (low resistance grounded—TN-S or high resistance grounded—IT) on the reliability and fault-tolerance of the system?

1.3. Paper Structure

This paper is organized as follows. The first section presented an introduction, including the motivation for the research and the specific research questions. The second section addresses system-level criteria for BIPV installations. A comparison between the different electrical architectures is given in section three. The fourth section will translate the system-level criteria into specific converter requirements and discuss relevant fault-tolerance and safety aspects more in-depth. The last section concludes this paper.

2. System Criteria for BIPV

This section describes the important system criteria that relate to the electrical installation of BIPV systems. These criteria serve as key performance indicators (KPIs) to evaluate the different interconnection methods, as shown in Figure 1. The next section will evaluate the different methods based on the derived KPIs.

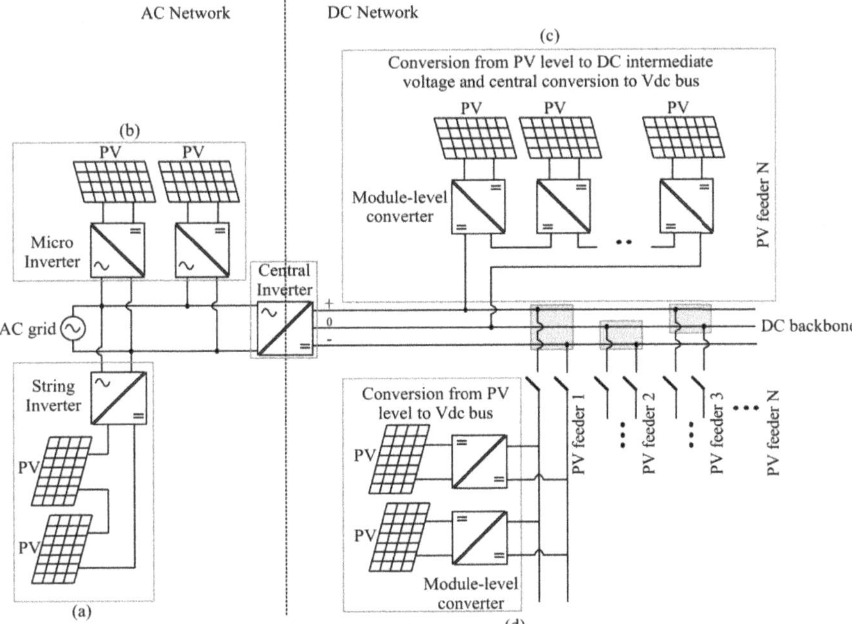

Figure 1. The possible electrical architectures for PV module interconnection to an AC or DC grid: (**a**) string inverters; (**b**) micro-inverters; (**c**) series power optimizer; and (**d**) parallel power optimizer.

2.1. Electrical Installation Methods

First, the different installation possibilities that can be used for the electrical installation of the BIPV system are briefly introduced. They are shown in Figure 1. In the string inverter approach (Figure 1a), several PV modules are series connected to form a string and are then coupled to the string inverter. Multiple strings are possible per inverter. The Maximum Power Point Tracking (MPPT) is done per string.

A second option is the use of micro-inverters (Figure 1b) where one inverter is coupled with a single PV module. The MPPT is done per module, and the output is coupled directly to the AC grid. A third option is the use of series power optimizers (Figure 1c). There is again one converter per module, but the output is now DC instead of AC. The output ports of multiple converters are then series connected and coupled to a central inverter that regulates the DC bus voltage. The last option is the use of parallel power optimizers (Figure 1d). As with the series power optimizers, their output is DC. However, the DC voltage level is higher. They are parallel connected to the same DC bus, and the MPPT is also done per module. The different installation possibilities will be further evaluated for BIPV applications in Section 2.8.

2.2. Energy Yield and Aesthetics

When designing the BIPV system, a trade-off needs to be made between the aesthetics and the energy-yield of the installation. A different approach exists between BAPV and BIPV. For BAPV, the electrical installation is typically done on an existing structure. A high energy yield is of primary importance as the motivation to place it can be found in reduced electricty costs. For BIPV, the energy yield is of secondary importance, whereas aesthetics and building regulation standards are primary factors to take into account [54]. Neglecting the importance of the energy yield could, however, lead to a less favourable position of BIPV compared to other measures that are used to reduce the energy consumption and ecological footprint of the building [55].

As compared to conventional PV systems, partial shading due to adjacent structures, building elements (e.g., pipes and ducts), frame edges, or the curvature of the installation is more prevalent in BIPV installations [36,47–49]. As a consequence, the power output of the shaded modules and, more importantly, the unshaded modules will be reduced, depending on the electrical configuration. The mismatch losses due to partial shading are specific for each installation, though studies indicate losses ranging from 5–25% [53,56–58].

This impact can, however, be limited by implementing the MPPT on smaller scales, e.g., per panel and not per string. To increase the power output, it is advisable to do the MPPT per module (distributed MPPT) and not per string. A micro-inverter (DC/AC) or power optimizer (DC/DC) solution is, thus, preferred to the traditional string inverter approach.

2.3. Flexibility

Flexibility relates to the architectural freedom during the design stage. In everyday structures, such as a casual office building, it can be advantageous to use arrays of standardized BIPV modules in order to reduce costs. This could, however, lead to a rather monotonous façade. When partial shading is not an issue, a string inverter could be used for the electrical installation, which is again beneficial from a cost point of view.

In more prestigious buildings, multiple types of PV (mono-, poly-, or thin film) can be present under a variety of forms. The electrical characteristics of these panels differ due to the different electrical properties of the used material, the size, and the orientation. Since aesthetics are of focal importance [54,55], this seems to be the preferred scenario, despite the higher impact on the electrical installation. Even when partial shading effects are negligible, the variety in output parameters favours the use of module-level converters as stringing becomes difficult or even impossible.

2.4. Modularity

This criterion relates to the ease of the practical construction of the installation. A modular system can easily be built and is the most plug-and-play option. This modularity aspect needs an evaluation for both the mechanical and electrical installation. The main focus of this paper, however, is on its electrical aspect. Modularity in its ultimate form would mean that the mechanical and electrical installation of the building can be carried out simultaneously by the same person. This person is preferably a construction worker with a minimal amount of extra training such that no skilled electrician is required to wire up the electrical installation after the façade itself has been placed. Relating to the previous aspects of flexibility, modularity does not mean that every BIPV module will be the same. The mechanical dimensions of a curtain wall BIPV module will be the same over a certain area of the building, e.g., one floor level, for the easiness of constructing the façade. Part of the BIPV module will be glass; the other part will be PV. The ratio glass/PV is, however, a parameter that can differ over different panels. This concept is illustrated in Figure 2 where BIPV modules are shown with 50% and 100% PV penetration. Furthermore, the glazing could also be changed to transparent PV modules which further increases the variety of electrical parameters in the system.

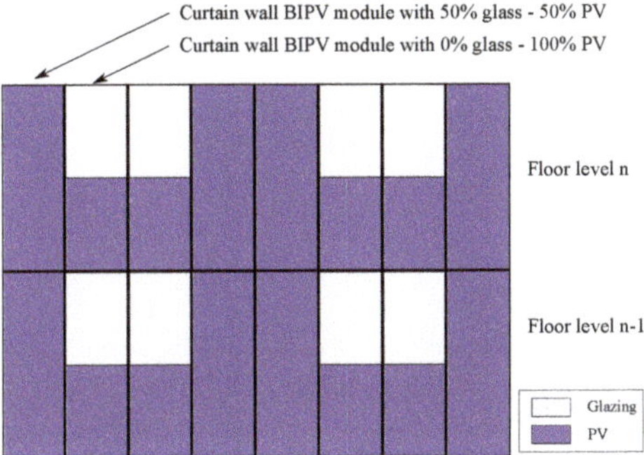

Figure 2. The zoom on a small part of a typical apartment or office building façade where building-integrated photovoltaic (BIPV) curtain wall modules can be used. To illustrate the modularity principle, the size of the curtain wall element remains the same but the ratio glass/photovoltaics (PV) can be changed.

Cabling is another factor that strongly influences the modularity of the system. Ideally, a preinstalled AC or DC bus structure would run inside the BIPV modules. By doing so, the BIPV modules can be directly interconnected at the moment of their installation. At the corners of the building, the cables of the BIPV modules of the same floor level can then be connected to the main AC or DC backbone.

2.5. Engineering Effort

The engineering effort and associated costs for the electrical installation should be minimized, leading to a lower overall system cost and a low threshold for architects to employ BIPV in their designs.

In BAPV, the engineering effort is already very low due to the standardization in panel sizes (standard 60 or 72 cells), the smaller system size, and the fact that cabling can be easily routed behind the panels.

A high degree of standardization and modularity in BIPV systems will reduce the engineering effort required to design the electrical system. Reference [43] already indicated that preinstalled MLCs in the BIPV modules are a promising solution to decrease costs. By doing so, all modules can be placed in parallel to a common bus structure. The different rated powers become a nonissue when a MLC is used, and the cabling from one module to the next can be preinstalled.

2.6. Reliability and Availability

As discussed in the previous points on the modularity and flexibility of the system, a fully integrated BIPV module with an integrated MLC is desired. The location of the MLC will be at the back of the PV panel or inside the metallic framework of the BIPV module. The major drawback of this choice is that the converter is difficult to reach after installation, making repairs or replacements undesirable or even impossible. The MLC lifetime should, thus, be comparable to the lifetime of the PV panel or even to the lifetime of the façade. The reliability of the converter becomes an important criterion to judge upon to enable this lifetime. A considerable amount of research is conducted to characterize the stresses of micro-inverters, considering mission profiles and degradation [59–61]. Commercial micro-inverter and power optimizers manufacturers use reliability as one of the key arguments to promote their products above competitors in the field and above the use of traditional string inverters. From Reference [62], the MTTF of string inverters is in the range of 20 years whereas micro-inverters report a MTTF of 300(+) years [63,64], certified by independent reliability test centers. Series power optimizers even report MTTF of 1000+ years, motivating that this is a consequence of the low amount of internal components, compared to string inverters [65].

No specific numbers have been found on parallel power optimizers. A higher reliability is, however, expected compared to micro-inverters as they only perform a DC/DC conversion, requiring less components and conversion steps. The converter reliability will, however, be lower compared to a series power optimizer, as a consequence of the higher voltage step-up, leading to more components and/or more complex circuit topologies [66,67]. For further calculations, the MTTF of parallel power optimizers is assumed to be 500 years.

From the above numbers, the converter reliability is plotted in Figure 3a for a time span of 40 years, corresponding to the desired lifetime of a BIPV façade. The string inverter performs the worst, leading to required replacements up to five times during the lifetime of a PV installation [59,68]. The replacements are not modeled in this study. The performance of micro-inverters and power optimizers is better, leading to an estimated reliability operational at 40 years of 87% for the micro-inverter, 96% for the series power optimizer (PO), and 92% for the parallel PO. These high numbers for distributed MPPT architectures have provoked skepticism within the PV industry and are criticized, as no or little field data is available to support these claims [62]. The numbers are mainly used to show specific trends, their numerical exactness is of secondary importance. Furthermore, the numbers that are reported in literature always referring to older converter designs that were installed several years ago. They can be used to derive a correct order of magnitude, but they do not take new designs with possible different failure mechanisms into account.

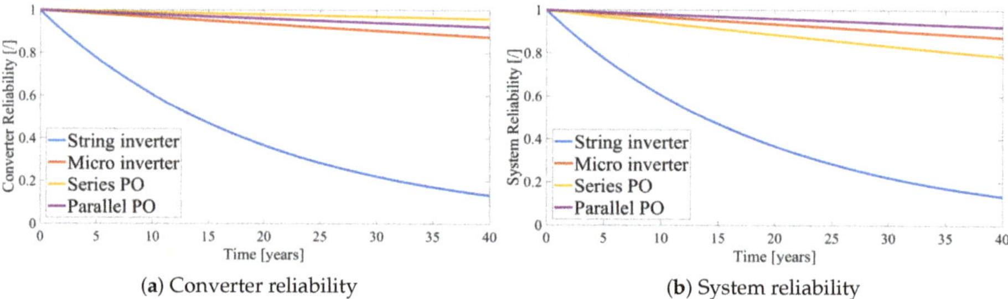

(a) Converter reliability   (b) System reliability

Figure 3. The predicted reliability of the different electrical architectures based on manufacturer reliability data. Note that repair is not modeled since the integrated converter is not servicable when installed in the module frame. The string inverter shows the lowest reliability but is installed on a servicable location which allows repair.

Even when special care is taken to have a reliable converter design, failures are inevitable. Hence, it is important that these failures only affect the converter and not the system as a whole. When a string inverter fails, the entire PV array becomes unavailable until the unit is repaired or replaced. As indicated in Reference [62], the independent and distributed architecture of MLC only affects one specific BIPV module. This leads to a reduced total power generation, but the system as a whole remains available. This is the case for micro-inverters and parallel power optimizers, if it is assumed that they are designed to fail safe. For series power optimizers, the malfunctioning of one device can lead to the malfunctioning of the entire string [69]. The string length of the leading commercial manufacturers is at least six and can go up to 25 (for a single phase grid connection). Taking these factors into account, the system reliability for the different cases is plotted in Figure 3b, showing that the parallel power optimizer approach leads to the highest system reliability. The reliability of the series power optimizer system has decreased due to the dependence on the correct functioning of the other optimizers within the string. Note that the MLC are assumed to be placed inside the metallic casing such that replacement is undesirable or even impossible. Specific failure modes of MLC and their consequences are investigated further on in this paper.

Furthermore, as reported by Reference [68], one of the main causes of PV inverter failure is related to inadequate protection from grid events, mainly surge voltages. To enhance the system's reliability, this factor needs to be taken into account.

2.7. Monitoring and Communication

To analyze and assess the performance of the system, it is recommended to collect data regarding the generated PV power. When this data is analyzed over a certain period of time, it can help to schedule maintenance for the installation (cleaning, testing, and a comparative analysis). Furthermore, peer-to-peer communication can be beneficial to detect fault scenarios, as will be discussed in detail in Section 4.

2.8. Technical Room Space and Cable Management—Case Study EnergyVille

The required technical room space (including technical shafts) for the BIPV installation needs to be minimal. This requirement is mainly valid for installations in densely populated areas where the price per square meter is elevated. Integrating the electronics and the cables is beneficial for this criterion. As already discussed under Section 2.4, using an AC or DC bus that runs along the modules can greatly simplify the cable management of the system.

To evaluate the technical room space, a case study has been carried out. Figure 4 shows the EnergyVille research institute building, located in Genk, Belgium. This building has a BAPV capacity of 369.15 kWp installed on the roof of the building, occupying a surface of 3326.17 m^2. The PV strings are connected to the grid via 23 string inverters of 8.5 and 20 kVA, requiring an inverter room of, approximately, 100 m^3.

The southern façade of this building has a strong potential for a façade BIPV installation due to its southern orientation and the absence of nearby obstacles. The available surface allows a peak power installation capacity of 97.8 kWp. Considering that each panel covers a surface of 1.63 m^2 and the available surface is around 531 m^2 (already disconsidering the occupied area by the windows) and taking into account that the partial shading deteriorates the energy generation by 10–13% [53], the estimated peak power of the southern façade is around 85 kWp. If a BIPV structure is installed and connected via local micro-converters, the installation leads to an increase of about 25% of the installed renewable generation while it does not require additional room space.

Figure 4. The EnergyVille building located in Genk, Belgium with 369.15 kW of peak PV installed on the roof.

3. Evaluation of the Electrical Installations for BIPV

The objective of this section is to give an overview of the possible installation methods to electrically interconnect BIPV modules. A discussion on the advantages and disadvantages of each system will be provided. The following criteria are considered: monitoring, modularity, engineering effort, immunity against AC disturbances, immunity against partial shading, and reliability. The interconnection of BIPV modules to the AC grid or DC backbone, shown in Figure 1, leads to several options that will be explored. Figure 1a,b show the interconnection of BIPV modules directly to the AC grid, while Figure 1c,d represents the possible configurations to interconnect the BIPV modules to a DC microgrid.

Regarding the connection from the PV voltage level to the AC grid (single or three-phase) string inverters are often used [70–72]. The BIPV modules are connected in series, and the string is connected to one inverter. Micro-inverters, where each PV panel has its own low power inverter, are another possibility to connect the panels directly to the AC grid [73–75].

To guarantee immunity against AC disturbances, the BIPV modules can be connected via DC power optimizers to an LVDC grid (bipolar or unipolar), as shown in Figure 1. In this type of grid connection, two options are highlighted: a series operation of power optimizers and a parallel operation of power optimizers.

The systems abovementioned have strengths and weaknesses that will be presented considering the predefined criteria in this paper for a BIPV electrical installation. Table 1 presents this overview.

3.1. String Inverters

In the string inverter configuration, the DC input voltage of the inverter is the series connection of the BIPV modules. This solution can lead to a lower energy yield during partial shading conditions, thereby degrading the overall system performance. The use of string inverters for BIPV installations is widely reported in the literature [58,76–86].

The analysis of Table 1 shows that string inverters have an inferior performance for all criteria that were established for BIPV installations. The system is not modular since stringing needs to be done, requiring a high engineering effort to optimally design the number and length of the PV strings for a given installation. Especially when different sizes and types of PV are used, stringing becomes extremely challenging. System reliability is the lowest of all four options, but this number might be misleading since no repair was modelled. This was done because MLCs are placed in locations that are difficult or impossible to reach after installation. However, the string inverter is placed inside the protected volume and can, thus, be repaired when a fault occurs.

3.2. Micro-Inverters

For the case of micro-inverters, the system improves over several aspects. The problem of partial shading is now solved due to the distributed MPPT. Micro-inverters allow the fitting inside the frame of the BIPV module, thus becoming a real MLC. This combination is commonly referred to as an AC module. Micro-inverters are still coupled to the AC grid, making them vulnerable to AC grid disturbances. This is a major drawback given that the MLC is not difficult to reach after installation. An overvoltage on the AC grid could, thus, lead to a shutdown of the entire BIPV installation without the possibility to repair. From a reliability point of view, the major drawback of micro-inverters is probably the large electrolytic capacitors that are required to buffer the power output ripple at twice the switching frequency [44].

In literature, only one example is found that reports the use of micro-inverters. This is the case for the Copenhagen International School in Denmark [87].

3.3. Power Optimizers

Compared to inverters, power optimizers (POs) benefit from an immunity against AC grid disturbances. They are similar to micro-inverters regarding the distributed MPPT approach but they do a DC/DC and not an AC/DC conversion. Because of this, their internal power electronics circuit topology can be simpler. This is especially true for series power optimizers (SPOs) of which the outputs are connected in series to obtain a high voltage at the end of the string. Their output voltage is currently controlled via an extra current source converter, which can be placed inside the building. The largest benefit of this approach is that their circuit topology can be a standard buck-boost converter with a low amount of required components and low voltage stresses on the components due to their lower output voltage [88–90]. Both factors favour the converter reliability and compactness of SPOs. As indicated in Table 1, SPOs have the same score as string inverters from a modularity and engineering effort point of view. This is because the installation is not necessarily plug-and-play. For small installations this might be relatively straightforward, assuming that, for example, one string per floor level is required and all BIPV modules can be simply put in series. For larger buildings with longer strings, this is not necessarily the case as the amount of SPOs per string is limited by the technology and correct cabling becomes challenging again. Furthermore, one malfunctioning converter can lead to the improper operation of the entire string, as discussed in Section 2.6. In the literature, several detection strategies are proposed to detect and overcome this issue [91,92].

Parallel POs (PPOs) show the best overall performance for use in BIPV electrical installations. As with SPOs, PPOs employ a DC/DC conversion, but all the outputs are placed in parallel to a common DC bus as shown in Figure 1d. The DC bus is controlled by a central voltage source converter which could be the same inverter that controls the LVDC micro grid.

PPOs combine the advantages of a distributed MPPT, a lower amount of internal components compared to inverters, a high degree of flexibility and modularity, no impact of AC grid disturbances, and the highest overall system reliability due to the independence of other converters.

To the author's best knowledge, no papers have been published where the SPO approach is used for BIPV applications. The use of PPOs for BIPV has first been reported by Reference [43], where a 200-V DC bus was used. The same approach was adopted by References [93–95], but a 380-V DC bus was employed. The reliability aspects concerning the embedment of this converter in the frame of a BIPV curtain wall element were treated in Reference [96].

Table 1. The qualitative and quantitative analysis of possible BIPV electrical installation architectures. The grey boxes indicate the preferred option.

System Network	Criteria						
	Monitoring	Modularity	Engineering Effort	Immunity against AC Disturbances	Flexibility	Immunity against Partial Shading	Predicted System Reliability after 40 Years
SI	no	low	high	no	low	no	13.5%
MI	yes	high	low	no	high	yes	87.5%
SPO	yes	low	high	yes	high	yes	78.9%
PPO	yes	high	low	yes	high	yes	92.3%

SI: String Inverter; MI: micro-inverter; SPO: Series Power Optimizer; PPO: Parallel Power Optimizer.

3.4. LVDC Grid

A low-voltage DC architecture is adopted because of three main arguments. Primarily, an LVDC grid has a higher compatibility with the DC devices in the system. PV systems, energy storage systems, electric vehicles, LED lighting, IT equipment, and drives operate natively on DC or require DC along the power conversion chain [97–101]. Consequently, an LVDC grid can simplify the conversion steps, which is not only limited to DC-DC power optimizers. That results in an increased efficiency, a reduced component part count, and an enhanced component-level reliability. Secondly, an LVDC grid can transmit more power provided the same copper conductor cross section, thereby lowering costs [99,101]. Thirdly, because power converters are predominantly present in LVDC grids, power flows are actively controllable as compared to passive rectifiers used in AC systems.

Compared to the previous arguments that hold for LVDC grids in general, what are the specific advantages of an LVDC grid in a BIPV context? At first, the fewer conversion steps lead to a lower component count, which leads to an inherent higher reliability and an increased power density. This is beneficial for a frame-integration of the modules. Secondly, since both the input and output power are DC quantities, no energy buffering is required to filter out the power pulsation occurring at twice the grid frequency. The required buffer capacitance is given by Equation (1). To increase the power density and decrease costs, this is usually an electrolytic capacitor. The long-term reliability of this component is often questioned and is, therefore, a distinct advantage if electrolytic capacitors can be left out of the design [44,102].

$$C_{DC} = \frac{P_{MPP}}{2 \cdot \omega_{grid} \cdot U_{DC} \cdot u_{ripple}} \tag{1}$$

The voltage level can be seen as a degree of freedom which allows for a balance of the converter and cabling costs in the system. A higher voltage level leads to smaller cable cross sections and, thus, to cheaper cabling. However, a lower voltage level can reduce the required gain of the converter and allows the employment of components with a lower voltage rating, which might lead to cost reductions for the converter. To date, there are no internationally recognized standards or agreements on the exact voltage level of future LVDC grids. In telecom applications, a unipolar 48 V grid has been used for decades. In datacenters, a transition is going on from AC to bipolar DC with a voltage level of 380 V (+190 V/0/−190 V) [103]. Rodriguez et al. proposed the use of a bipolar 1500 V (+750 V/0/−750 V) grid for LVDC distribution that can be further divided in a bipolar 750 V (+375 V/0/−375 V) for high power loads and a unipolar 48 V bus for low power loads [104].

4. Converter Requirements

The discussion in Section 2.8 has led to the insight that a parallel power optimizer (PPO) approach is the most appropriate method to tackle the requirements that were set forward in Section 2. However, the previous analysis has shown that up to this date, the majority of BIPV installations still employ the more traditional string inverters to interconnect the BIPV modules. This section focuses on the specific requirements of a PPO for use in BIPV modules as an MLC. Every MLC is connected in parallel to the LVDC grid (the common DC bus) as shown in Figure 1d.

4.1. Compactness

The structural strength of BIPV modules is achieved via extruded aluminium profiles, and the MLC is preferably installed inside these cavities. In BAPV modules, the MLC is installed at the back of the PV panel, but in the case of BIPV, this space is occupied by thermal insulation material.

In general, a flat and long design is preferred to fit inside the framework. Currently commercially available MLC converters do not have this form factor, making it impossible to fit in this cavity. The implementation of wide-bandgap components such as Silicon Carbide (Sic) or Gallium Nitride (GaN) in the converter topology can help to achieve the required power density goals by an increased switching frequency, leading to smaller passive devices. The avoidance of transformers or inductors can lead to further size reductions, usually accomplished via switched capacitor circuits [105]. However, switched capacitor circuits can suffer from a lower reliability given that capacitors are one of the dominant components that lead to device failures [59]. Especially when MultiLayer Ceramic Capacitors (MLCCs) are used, a careful design is important due to their dominant short-circuit failure mode [102].

4.2. Wide Power and Input Voltage Range

In order to meet the scalability and flexibility requirements, it is preferable that the converter is compatible with a wide variety of PV panel types (mono-, polycristalline, and thin film) and sizes. The type of panel will already dictate part of its electrical characterisitics, but this is also influenced by the active surface, leading to a higher or lower amount of active cells. Apart from the standard 60 or 72 cell panel size, strongly deviating designs can occur, depending on the architect's needs. Currently, commercial and research projects still focus on standard 60 or 72 cell PV modules, leading to designs that work for 20–50 V input voltage. This voltage range needs to be widened on both ends for future converter designs. The low input voltages might, however, lead to lower efficiencies due to the higher step-up [66,67]. Therefore, BIPV modules with a higher output voltage are more attractive from a power electronics point of view. This allows for simpler and smaller topologies to be used.

Furthermore, to optimize the energy-yield of the installation, the converters need to cover a wide input power range, preferably from 10–100% of the nominal peak power. The challenge is to maintain a decent efficiency over the entire operating range as typically the efficiency curve strongly decreases on the low-power operating range. Interleaved converters can help to achieve this goal by controlling the amount of phases based on the input power, thereby allowing a flatter efficiency curve [106].

4.3. Temperature Range and Cooling

Experimental results from previous studies indicate that the temperature in and around façade BIPV panels can strongly exceed 100 °C, depending on the installation and type of grid ventilation [53]. Measurements inside the framework where the converter will be placed indicate temperatures around 80 °C, leading to high thermal stresses for the converter components. As temperature and temperature cycling are one of the major stressors for power electronic components [59], cooling or heat-spreading will be required to keep the component temperature low enough to meet the reliability requirements.

However, the physical dimensions of the heat sink need to be limited such that the size does not conflict with the compactness requirements. An alternative gaining importance in space-constrained applications is evaporative cooling using ultrathin heat pipes [107]. Active cooling using forced air flow is not preferred as the lifetime of fans is limited due to wear-out of the bearings [62], whereas active liquid cooling is considered to be expensive and difficult to implement in the system. BIPVT systems, where the generated heat of the PV panel is partly recovered in the building via forced convection, seems to gain lots of research interest [12]. No studies have been carried out that check whether this is also an effective approach to cool integrated electronics such as the MLC.

4.4. Lifetime

As already discussed in Section 2.6, the lifetime of the converter should be comparable to the lifetime of the PV panel, which is currently in the order of 25 years with a warranted power output of 80% or more. This is, however, still not equal to the current lifetime of a façade in the order of 30 to 50 years and which is also the design target for BIPV façades.

Micro-inverter and power optimizer producers provide warranties up to 25 years for their products [64,69]. These warranties are valid for converters that are attached to the back of a BAPV module and not in the framework of a BIPV module, which is a more challenging environment.

In Reference [108], an experimental study with accelerated lifetime tests was conducted to determine the reliability issues of module-level power electronics converters. Their conclusion states that a significant number of devices failed due to the applied stress and that this is most likely a consequence of design issues rather than manufacturing issues. Keeping this information in mind, combined with more severe temperature stresses that are expected in BIPV compared to BAPV, a design-for-reliability approach for BIPV MLCs is required.

4.5. Fail-Safe Functionalities

Fault-tolerant or fail-safe requirements relate to the correct functioning of the electrical system, after the occurrence of faults. When one MLC fails, it is important that this converter gets isolated from the system such that the BIPV feeder can remain operational with the remaining converters. This aids in achieving the availability target and is beneficial for the Return On Investment (ROI) of the installation [44]. Failures of PV arrays, their origin, consequences, and mitigation techniques are provided in Reference [109], but the failure of a single PV module connected via a power optimizer to an LVDC grid has not been covered. As the fault and leakage currents are much lower when the fault occurs on the PV side, this situation is different from a failure in a PV array. Novel fault detection methods as proposed in Reference [110] are recommended here.

In general, two types of faults can be distinguished: earth faults and short-circuit faults. This is shown in Figure 5. The influence on the availability of the system will be discussed for the case of earth faults, as they are related to the LVDC grid configuration and the presence of a high frequency transformer.

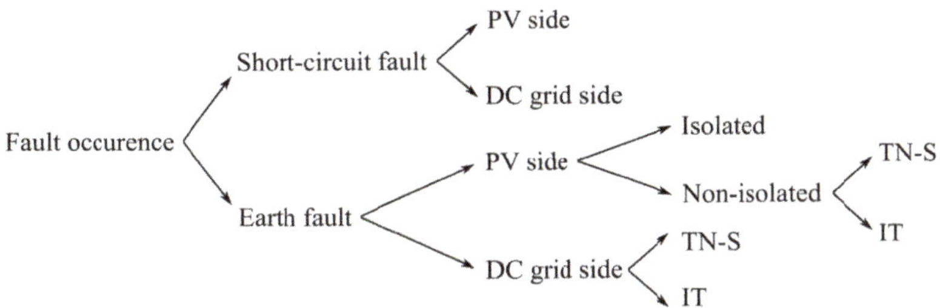

Figure 5. An overview of the possible faults in a fault tree structure. A distinction is made between short-circuit and earth faults and the input (PV) versus the output (DC grid) side of the converter.

4.5.1. Earth Fault Selectivity

Assume an earth fault occurs at the output terminal of the MLC. The required response will depend on the LVDC grid configuration. In an IT grid, it is possible to organize the protection centrally, for example, as an extra feature of the central AC/DC converter. If all the protection equipment and logic is contained within this central box, communication is necessary between the central converter and the MLCs. When an earth fault is detected, the central converter can force a sequential shutdown of the MLCs in order to localize and isolate the faulty converter. This specific strategy is only possible in the case of an IT grid configuration, as it allows a certain time to react and search the earth fault. It comes, however, at the cost of a more complex protection scheme, based on communication between the MLCs and a central controller. The MLC needs to be equipped with a relay such that it can entirely disconnect from the LVDC grid when required. The central inverter needs to be equipped with an isolation monitoring device (IMD). Due to the extra costs and increased control complexity, this protection scheme is not recommended. The situation is, however, different for a TN-S grid configuration. When an earth fault occurs, the situation is similar to a short-circuit fault. The earth fault protection can, thus, be identical to the local short circuit protection measures. This can be, for example, a fuse. The fault current will be determined by the amount of PV generation of the other converters, the total capacitance of the LVDC grid, and the fault impedance.

4.5.2. Galvanic Isolation: Yes or No?

A transformer is often regarded as a component that reduces the overall efficiency and strongly increases the cost and volume of a DC/DC converter. It has, however, been successfully implemented in several designs [43,111,112] to realize the high step-up from the PV voltage level to the LVDC voltage level. In this work, an extra advantage of galvanic isolation will be analyzed.

As can be seen from the fault tree in Figure 5, the presence of a transformer affects only the situation where an earth fault occurs on the PV side. The three possible situations for this fault (assuming PV+ to PE) are analyzed in Table 2. A similar analysis is possible for a PV− to PE fault, but it is omitted here as the outcome is the same. This table provides evidence that a (high frequency) transformer in between the PV terminals and the LVDC grid improves the fault-tolerance and, thus, the reliability of the system by creating a very local IT grid. As in an IT system, a first fault has no direct consequence but merely places the potentials on the same level. If the converter has a local isolation monitoring device installed, it can sense this fault and shut down, but this is not a strict requirement. A second fault will instantly lead to a short-circuit at the PV terminals, which is measurable and requires a converter shut down and disconnection from the LVDC grid.

Table 2. An overview on the converter PV side earth faults for different grid configurations and distinguishing galvanic isolations.

Fault Schematic	Grid Type	Isolated	Consequence	Required Response	Detection and Protection Device
	NA	Yes	No direct consequence	The converter can remain operational. A second fault on the PV side will result in a short-circuit between PV+ and PV−.	/
	TN-S	No	Short-circuit current between PV+ and PV− determined by PV panel I-V characteristic	Shut down and disconnect converter from input and output	Local fault or insulation monitoring device
	IT	No	Short-circuit dependent on the grouinding impedanc Z_G	Shut down and disconnect converter from input and output	Local fault or insulation monitoring device

5. Conclusions

BIPV modules are installed during, not after, the construction phase and have a profound impact on the electrical installation and construction planning of a building. In this paper, system criteria were established for the electrical installations of building-integrated photovoltaics. These criteria serve as key performance indicators. Apart from energy-yield, factors such as engineering effort, availability, and modularity must also be considered when designing the installation as they impact project installation costs. Currently available electrical installation architectures, such as string inverters, micro-inverters, and series and parallel power optimizers, were compared according to the aforementioned criteria, favoring parallel module-level converters connected to a low-voltage DC grid for BIPV applications. LVDC grids allow a further reduction of the costs and an increase in the power density and lifetime compared to traditional AC grids. This is mainly because of the lower amount of conversion stages, leading to a lower amount of components. Although, the use of string inverters is mostly reported for BIPV. The requirements for BIPV module-level converters mainly differ from regular PV converters in terms of compactness, input range, and operating temperature range. Several methods were discussed that allow an improvement of these aspects for future BIPV converter designs. Furthermore, BIPV module-level converters must incorporate fault-tolerance techniques to meet the building element's lifetime requirements. Due to the difficulties in replacing the converter when it is embedded inside the framework of the BIPV module, it is of utmost importance that the converter is designed to fail safe such that a converter failure does not result in a system failure. Fault conditions in BIPV module-level converters were considered for different grounding configurations. Compared to the current trend of going to non-isolated PV converters, we recommend the use of transformer-isolated topologies which increase the fault tolerance of the installation as a whole. The preferred grounding configuration is TN-S, as it allows for simple fault discrimination based on the current intensity.

Author Contributions: Conceptualization, S.R.; resources, J.D.; supervision, J.D.; validation, M.D.V. and G.V.d.B.; writing—original draft, S.R.; writing—review and editing, M.D.V., G.V.d.B., and J.D.

Funding: This project received the support of the European Union, the European Regional Development Fund ERDF, Flanders Innovation & Entrepreneurship, and the Province of Limburg. G. Van den Broeck, funded by a PhD grant of the Research Foundation Flanders (FWO).

Conflicts of Interest: The authors declare no conflict of interest.

Abbreviations

The following abbreviations are used in this manuscript:

AC	Alternating Current
BAPV	Building-Applied Photovoltaics
BIPV	Building-Integrated Photovoltaics
BIPVT	Building-Integrated Photovoltaics Thermal
DC	Direct Current
EPBD	Energy Performance of Buildings Directive
GaN	Gallium Nitirde
IMD	Insulation Monitoring Device
IT	High-resistance grounded (from French: Isolé—Terre)
KPI	Key Performance Indicator
LVDC	Low-Voltage Direct Current
MI	Micro-inverter
MLC	Module Level Converter
MLCC	Multilayer Ceramic Capacitor
MPP	Maximum Power Point

MPPT	Maximum Power Point Tracker
NZEB	Near Zero Energy Buildings
PPO	Parallel Power Optimizer
PV	Photovoltaics
SET	Strategic Energy Technology
SI	String Inverter
SiC	Silicon Carbide
SPO	Series Power Optimizer
TN-S	Low-resistance grounded (from French: Terre-Neutre Séparé)

References

1. Sinapis, K.; van den Donker, M. *BIPV REPORT 2013—State of the Art in Building Integrated Photovoltaics*; Technical Report; Solar Energy Application Centre: Eindhoven, The Netherlands, 2013.
2. Prasad, D.; Snow, M. *Designing with Solar Power: A Source Book for Building Integrated Photovoltaics (BiPV)*; Routledge: Abingdon-on-Thames, UK, 2014.
3. Verberne, G.; Bonomo, P.; Frontini, F. BIPV Products for Facades and Roofs: A Market Analysis. In Proceedings of the European Photovoltaic Solar Energy Conference and Exhibition, Amsterdam, The Netherlands, 22–26 September 2014.
4. Fath, K.; Stengel, J.; Sprenger, W.; Wilson, H.R.; Schultmann, F.; Kuhn, T.E. A method for predicting the economic potential of (building-integrated) photovoltaics in urban areas based on hourly Radiance simulations. *Sol. Energy* **2015**, *116*, 357–370. [CrossRef]
5. Shah, A.V.; Schade, H.; Vanecek, M.; Meier, J.; Vallat-Sauvain, E.; Wyrsch, N.; Kroll, U.; Droz, C.; Bailat, J. Thin-film silicon solar cell technology. *Prog. Photovolt. Res. Appl.* **2004**, *12*, 113–142. [CrossRef]
6. Hegedus, S. Thin film solar modules: The low cost, high throughput and versatile alternative to Si wafers. *Prog. Photovolt. Res. Appl.* **2006**, *14*, 393–411. [CrossRef]
7. European Commision. *Towards an Integrated Strategic Energy Technology (SET) Plan: Accelerating the European Energy System Transformation*; European Commision: Brussels, Belgium, 2015.
8. EU. Directive 2010/31/EU of the European Parliament and of the Council of 19 May 2010 on the energy performance of buildings (recast). *Off. J. Eur. Union* **2010**, *3*, 13–35. [CrossRef]
9. Osseweijer, F.J.; van den Hurk, L.B.; Teunissen, E.J.; van Sark, W.G. A comparative review of building integrated photovoltaics ecosystems in selected European countries. *Renew. Sustain. Energy Rev.* **2018**, *90*, 1027–1040. [CrossRef]
10. Defaix, P.R.; van Sark, W.G.; Worrell, E.; de Visser, E. Technical potential for photovoltaics on buildings in the EU-27. *Sol. Energy* **2012**, *86*, 2644–2653. [CrossRef]
11. Kamel, R.; Ekrami, N.; Dash, P.; Fung, A.; Hailu, G. BIPV/T+ASHP: Technologies for NZEBs. *Energy Procedia* **2015**, *78*, 424–429. [CrossRef]
12. Biyik, E.; Araz, M.; Hepbasli, A.; Shahrestani, M.; Yao, R.; Shao, L.; Essah, E.; Oliveira, A.C.; del Caño, T.; Rico, E.; et al. A key review of building integrated photovoltaic (BIPV) systems. *Eng. Sci. Technol. Int. J.* **2017**, *20*, 833–858. [CrossRef]
13. Delisle, V.; Kummert, M. Cost-benefit analysis of integrating BIPV-T air systems into energy-efficient homes. *Sol. Energy* **2016**, *136*, 385–400. [CrossRef]
14. Zhang, T.; Wang, M.; Yang, H. A Review of the Energy Performance and Life-Cycle Assessment of Building-Integrated Photovoltaic (BIPV) Systems. *Energies* **2018**, *11*, 3157. [CrossRef]
15. Park, J.; Hengevoss, D.; Wittkopf, S. Industrial Data-Based Life Cycle Assessment of Architecturally Integrated Glass-Glass Photovoltaics. *Buildings* **2018**, *9*, 8. [CrossRef]
16. Xu, X.; Feng, G.; Chi, D.; Liu, M.; Dou, B. Optimization of Performance Parameter Design and Energy Use Prediction for Nearly Zero Energy Buildings. *Energies* **2018**, *11*, 3252. [CrossRef]

17. Ballarini, I.; De Luca, G.; Paragamyan, A.; Pellegrino, A.; Corrado, V. Transformation of an Office Building into a Nearly Zero Energy Building (nZEB): Implications for Thermal and Visual Comfort and Energy Performance. *Energies* **2019**, *12*, 895. [CrossRef]
18. Martín-Chivelet, N.; Gutiérrez, J.C.; Alonso-Abella, M.; Chenlo, F.; Cuenca, J. Building retrofit with photovoltaics: Construction and performance of a BIPV ventilated façade. *Energies* **2018**, *11*, 1719. [CrossRef]
19. Corrao, R. Mechanical tests on innovative BIPV façade components for energy, seismic, and aesthetic renovation of high-rise buildings. *Sustainability* **2018**, *10*, 4523. [CrossRef]
20. Martinopoulos, G.; Serasidou, A.; Antoniadou, P.; Papadopoulos, A.M. Building integrated shading and building applied photovoltaic system assessment in the energy performance and thermal comfort of office buildings. *Sustainability* **2018**, *10*, 4670. [CrossRef]
21. Goncalves, J.; Lehmann, J.; Yordanov, G.; Parys, W.; Baert, K.; Saelens, D. Experimental Performance of a Curtain Wall BIPV Element Under Realistic Boundary Conditions. In Proceedings of the European Photovoltaic Solar Energy Conference (EUPVSEC 2018), Brussels, Belgium, 24–28 September 2018.
22. Goncalves, J.; Spiliotis, K.; Lehmann, J.; Baert, K.; Driesen, J.; Saelens, D. Experimental validation of a BIPV curtain wall model for building energy simulations. In Proceedings of the Advanced Building Skins Conference (ABS 2018), Bern, Switzerland, 1–2 October 2018.
23. Lehmann, J.; Parys, W.; Goncalves, J.; Baert, K.; Saelens, D. Experimental analysis of the performance of a BIPV curtain wall component. In Proceedings of the Advanced Building Skins Conference (ABS 2017), Bern, Switzerland, 2–3 October 2017.
24. Escarre, J.; Li, H.Y.; Sansonnens, L.; Galliano, F.; Cattaneo, G.; Heinstein, P.; Nicolay, S.; Bailat, J.; Eberhard, S.; Ballif, C.; et al. When PV modules are becoming real building elements: White solar module, a revolution for BIPV. In Proceedings of the 2015 IEEE 42nd Photovoltaic Specialist Conference, PVSC 2015, New Orleans, LA, USA, 14–19 June 2015; pp. 1–2. [CrossRef]
25. Yu, H.; Wang, Q.; Lu, C.; Wei, C. The research on a new type of BIPV modules constructed by thin-film photovoltaic panel(or module)/PU/color organic-coated steel plate. In Proceedings of the 2014 IEEE 40th Photovoltaic Specialist Conference, PVSC 2014, Denver, CO, USA, 8–13 June 2014; pp. 2724–2727. [CrossRef]
26. Martellotta, F.; Cannavale, A.; Ayr, U. Comparing energy performance of different semi-transparent, building-integrated photovoltaic cells applied to "reference" buildings. *Energy Procedia* **2017**, *126*, 219–226. [CrossRef]
27. Virtuani, A.; Strepparava, D. Modelling the performance of amorphous and crystalline silicon in different typologies of building-integrated photovoltaic (BIPV) conditions. *Sol. Energy* **2017**, *146*, 113–118. [CrossRef]
28. Bonomo, P.; Saretta, E.; Frontini, F. Towards the implementation of a BIM-based approach in BIPV sector. In Proceedings of the Advanced Building Skins Conference, Bern, Switzerland, 1–2 October 2018.
29. Alamy, P.; Nguyen, V.K. PVSITES software tools for BIM-based design and simulation of BIPV systems. In Proceedings of the Advanced Building Skins Conference, Bern, Switzerland, 1–2 October 2018.
30. Jakica, N. State-of-the-art review of solar design tools and methods for assessing daylighting and solar potential for building-integrated photovoltaics. *Renew. Sustain. Energy Rev.* **2018**, *81*, 1296–1328. [CrossRef]
31. Fazelpour, F.; Ziasistani, N.; Nazari, P.; Nazari, P.; Ziasistani, N. Techno-Economic Assessment of BIPV Systems in Three Cities of Iran. *Proceedings* **2018**, *2*, 1474. [CrossRef]
32. Zhang, Y.; Yan, Z.; Li, L.; Yao, J. A Hybrid Building Power Distribution System in Consideration of Supply and Demand-Side: A Short Overview and a Case Study. *Energies* **2018**, *11*, 3082. [CrossRef]
33. Elinwa, U.K.; Radmehr, M.; Ogbeba, J.E. Alternative energy solutions using BIPV in apartment buildings of developing countries: A case study of North Cyprus. *Sustainability* **2017**, *9*, 1414. [CrossRef]
34. An, H.J.; Yoon, J.H.; An, Y.S.; Heo, E. Heating and cooling performance of office buildings with a-Si BIPV windows considering operating conditions in temperate climates: The case of Korea. *Sustainability* **2018**, *10*, 4856. [CrossRef]
35. Kuo, H.J.; Hsieh, S.H.; Guo, R.C.; Chan, C.C. A verification study for energy analysis of BIPV buildings with BIM. *Energy Build.* **2016**, *130*, 676–691. [CrossRef]

36. Cerón, I.; Caamaño-Martín, E.; Neila, F.J. State-of-the-art of building integrated photovoltaic products. *Renew. Energy* **2013**, *58*, 127–133. [CrossRef]

37. Montoro, D.F.; Vanbuggenhout, P.; Ciesielska, J. *Building Integrated Photovoltaics: An Overview of the Existing Products and Their Fields of Application*; Technical Report; Prepared in the Framework of the European Funded Project; SUNRISE: Saskatoon, SK, Canada, 2011.

38. Zanetti, I.; Bonomo, P.; Frontini, F.; Saretta, E.; van den Donker, M.; Vossen, F.; Folkerts, W. *Building Integrated Photovoltaics: Product Overview for Solar Building Skins- Status Report 2017*; Technical Report; Swiss BIPV Comptence Centre (SUPSI): Mano, Switzeland; Solar Energy Application Centre (SEAC): Eindhoven, The Netherlands, 2017.

39. Vasiliev, M.; Nur-E-Alam, M.; Alameh, K. Recent Developments in Solar Energy-Harvesting Technologies for Building Integration and Distributed Energy Generation. *Energies* **2019**, *12*, 1080. [CrossRef]

40. Pearce, J.; Meldrum, J.; Osborne, N. Design of Post-Consumer Modification of Standard Solar Modules to Form Large-Area Building-Integrated Photovoltaic Roof Slates. *Designs* **2017**, *1*, 9. [CrossRef]

41. Attoye, D.E.; Aoul, K.A.T.; Hassan, A. A review on building integrated photovoltaic façade customization potentials. *Sustainability* **2017**, *9*, 2287. [CrossRef]

42. Asfour, O.S. Solar and shading potential of different configurations of building integrated photovoltaics used as shading devices considering hot climatic conditions. *Sustainability* **2018**, *10*, 4373. [CrossRef]

43. Liu, B.; Duan, S.; Cai, T. Photovoltaic DC-building-module-based BIPV system-concept and design considerations. *IEEE Trans. Power Electron.* **2011**, *26*, 1418–1429. [CrossRef]

44. Formica, T.J.; Khan, H.A.; Pecht, M.G. The Effect of Inverter Failures on the Return on Investment of Solar Photovoltaic Systems. *IEEE Access* **2017**, *5*, 21336–21343. [CrossRef]

45. Agamy, M.S.; Harfman-Todorovic, M.; Elasser, A.; Chi, S.; Steigerwald, R.L.; Sabate, J.A.; McCann, A.J.; Zhang, L.; Mueller, F.J. An Efficient Partial Power Processing DC/DC Converter for Distributed PV Architectures. *IEEE Trans. Power Electron.* **2014**, *29*, 674–686. [CrossRef]

46. Dragicevic, T.; Lu, X.; Vasquez, J.C.; Guerrero, J.M. DC Microgrids—Part 1: A Review of Control Strategies and Stabilization Techniques. *IEEE Trans. Power Electron.* **2016**, *31*, 4876–4891. [CrossRef]

47. Petter Jelle, B.; Breivik, C.; Drolsum Røkenes, H. Building integrated photovoltaic products: A state-of-the-art review and future research opportunities. *Sol. Energy Mater. Sol. Cells* **2012**, *100*, 69–96. [CrossRef]

48. Heinstein, P.; Ballif, C.; Perret-Aebi, L.E. Building integrated photovoltaics (BIPV): Review, potentials, barriers and myths. *Green* **2013**, *3*, 125–156. [CrossRef]

49. Ikkurti, H.P.; Saha, S. A comprehensive techno-economic review of microinverters for Building Integrated Photovoltaics (BIPV). *Renew. Sustain. Energy Rev.* **2015**, *47*, 997–1006. [CrossRef]

50. Bidram, A.; Davoudi, A.; Balog, R.S. Control and circuit techniques to mitigate partial shading effects in photovoltaic arrays. *IEEE J. Photovolt.* **2012**, *2*, 532–546. [CrossRef]

51. Freitas, S.; Brito, M. Maximizing the Solar Photovoltaic Yield in Different Building Facade Layouts. In Proceedings of the European Photovoltaic Solar Energy Conference and Exhibition, Hamburg, Germany, 14–18 September 2015; pp. 1–5. [CrossRef]

52. Sinapis, K.; Litjens, G.; van den Donker, M.; Folkerts, W.; van Sark, W. Outdoor characterization and comparison of string and MLPE under clear and partially shaded conditions. *Energy Sci. Eng.* **2015**, *3*, 510–519. [CrossRef]

53. Broeck, G.V.D.; Parys, W.; Goverde, H.; Putten, S.V.D.; Poortmans, J.; Driesen, J.; Baert, K. Experimental analysis of the performance of facade-integrated BIPV in different configurations. In Proceedings of the 32nd European Photovoltaic Solar Energy Conference and Exhibition, Munich, Germany, 20–24 June 2016; pp. 1–5.

54. Marc, M. Towards a broad BIPV deployment: What is needed? In Proceedings of the EnergyVille BIPV Workshop, Genk, Belgium, 8 December 2015.

55. Saelens, D. EnergyVille BIPV Roadmap. In Proceedings of the Belgian BIPV Workshop, Brussels, Belgium, 7 September 2017.

56. Gross, M.A.; Martin, S.O.; Pearsall, N.M. Estimation of output enhancement of a partially shaded BIPV array by the use of AC modules. In Proceedings of the Conference Record of the Twenty Sixth IEEE Photovoltaic Specialists Conference, Anaheim, CA, USA, 29 September–3 October 1997; pp. 1381–1384. [CrossRef]

57. Van der Borg, N.; Jansen, M. Energy loss due to shading in a BIPV application. In Proceedings of the 3rd World Conference on Photovoltaic Energy Conversion, Osaka, Japan, 11–18 May 2003; Volume 3, pp. 2220–2222.

58. Zomer, C.; Nobre, A.; Cassatella, P.; Reindl, T.; Rüther, R. The balance between aesthetics and performance in building-integrated photovoltaics in the tropics. *Prog. Photovolt. Res. Appl.* **2014**, *22*, 744–756. [CrossRef]

59. Wang, H.; Liserre, M.; Blaabjerg, F. Toward reliable power electronics: Challenges, design tools, and opportunities. *IEEE Ind. Electron. Mag.* **2013**, *7*, 17–26. [CrossRef]

60. Shen, Y.; Wang, H.; Blaabjerg, F. Reliability Oriented Design of a Grid-Connected Photovoltaic Microinverter. In Proceedings of the 2017 IEEE 3rd International Future Energy Electronics Conference and ECCE Asia, Kaohsiung, Taiwan, 3–7 June 2017; pp. 81–86.

61. Zare, M.H.; Mohamadian, M.; Wang, H.; Blaabjerg, F. Reliability assessment of single-phase gridconnected PV microinverters considering mission profile and uncertainties. In Proceedings of the 8th Power Electronics, Drive Systems and Technologies Conference, PEDSTC 2017, Mashhad, Iran, 14–16 February 2017; pp. 377–382. [CrossRef]

62. Chung, H.S.H.; Wang, H.; Blaabjerg, F.; Pecht, M. *Reliability of Power Electronic Converter Systems*; Institution of Engineering and Technology: Stevenage, UK, 2016; p. 489. [CrossRef]

63. ABB Group. *Application Note on micro-inverter Reliability*; Technical Report; ABB Group: Zürich, Switzerland, 2015.

64. Enphase. *Reliability of Enphase Micro-Inverters*; Enphase White Paper; Enphase: Fremont, CA, USA, 2009; pp. 1–10.

65. Routledge, J. *Panel Level Power Electronics—From Vision to Commercial Reality*; Technical Report; Tigo Energy: Los Gatos, CA, USA, 2015.

66. Josias de Paula, W.; Oliveira Júnior, D.D.S.; Pereira, D.D.C.; Tofoli, F.L. Survey on non-isolated high-voltage step-up dc–dc topologies based on the boost converter. *IET Power Electron.* **2015**, *8*, 2044–2057. [CrossRef]

67. Forouzesh, M.; Siwakoti, Y.P.; Gorji, S.A.; Blaabjerg, F.; Lehman, B. Step-Up DC-DC converters: A comprehensive review of voltage-boosting techniques, topologies, and applications. *IEEE Trans. Power Electron.* **2017**, *32*, 9143–9178. [CrossRef]

68. Petrone, G.; Spagnuolo, G.; Teodorescu, R.; Veerachary, M.; Vitelli, M. Reliability Issues in Photovoltaic Power Processing Systems. *IEEE Trans. Ind. Electron.* **2008**, *55*, 2569–2580. [CrossRef]

69. SolarEdge. *Power Optimizers Installation Guide*; Technical Report; SolarEdge: Herzliya, Israel, 2016.

70. Boztepe, M.; Guinjoan, F.; Velasco-Quesada, G.; Silvestre, S.; Chouder, A.; Karatepe, E. Global MPPT Scheme for Photovoltaic String Inverters Based on Restricted Voltage Window Search Algorithm. *IEEE Trans. Ind. Electron.* **2014**, *61*, 3302–3312. [CrossRef]

71. Sangwongwanich, A.; Yang, Y.; Blaabjerg, F.; Sera, D. Delta power control strategy for multi-string grid-connected PV inverters. In Proceedings of the 2016 IEEE Energy Conversion Congress and Exposition (ECCE), Milwaukee, WI, USA, 18–22 September 2016; pp. 1–7. [CrossRef]

72. Zhao, X.; Zhang, L.; Born, R.; Lai, J.S. Solution of input double-line frequency ripple rejection for high-efficiency high-power density string inverter in photovoltaic application. In Proceedings of the 2016 IEEE Applied Power Electronics Conference and Exposition (APEC), Long Beach, CA, USA, 20–24 March 2016; pp. 1148–1154. [CrossRef]

73. Mohammadi, S.; Zarchi, H.A. An interleaved high-power two-switch flyback inverter with a fast and robust maximum power point tracker. In Proceedings of the 2016 7th Power Electronics and Drive Systems Technologies Conference (PEDSTC), Tehran, Iran, 16–18 February 2016; pp. 320–325. [CrossRef]

74. Melo, F.C.; Garcia, L.S.; Freitas, L.C.; Coelho, E.A.A.; Farias, V.J.; Freitas, L.C.G. Proposal of a Photovoltaic AC-Module with a Single-Stage Transformerless Grid-Connected Boost Microinverter (STBM). *IEEE Trans. Ind. Electron.* **2017**, *65*, 2289–2301. [CrossRef]

75. Zengin, S.; Deveci, F.; Boztepe, M. Decoupling Capacitor Selection in DCM Flyback PV Microinverters Considering Harmonic Distortion. *IEEE Trans. Power Electron.* **2013**, *28*, 816–825. [CrossRef]

76. Mun, H.; Ho, J.; Chul, S.; Shin, U.C. Operational power performance of south-facing vertical BIPV window system applied in office building. *Sol. Energy* **2017**, *145*, 66–77. [CrossRef]

77. Aste, N.; Pero, C.D.; Leonforte, F. The first Italian BIPV project: Case study and long-term performance analysis. *Sol. Energy* **2016**, *134*, 340–352. [CrossRef]

78. Pola, I.; Chianese, D.; Bernasconi, A. Flat roof integration of a-Si triple junction modules laminated together with flexible polyolefin membranes. *Sol. Energy* **2007**, *81*, 1144–1158. [CrossRef]

79. Fanni, L.; Virtuani, A.; Chianese, D. A detailed analysis of gains and losses of a fully-integrated flat roof amorphous silicon photovoltaic plant. *Sol. Energy* **2011**, *85*, 2360–2373. [CrossRef]

80. Eke, R.; Senturk, A. Monitoring the performance of single and triple junction amorphous silicon modules in two building integrated photovoltaic (BIPV) installations. *Appl. Energy* **2013**, *109*, 154–162. [CrossRef]

81. Pillai, R.; Aaditya, G.; Mani, M.; Ramamurthy, P. Cell (module) temperature regulated performance of a building integrated photovoltaic system in tropical conditions. *Renew. Energy* **2014**, *72*, 140–148. [CrossRef]

82. Essah, E.A.; Rodriguez, A.; Glover, N. Assessing the performance of a building integrated BP c-Si PV system. *Renew. Energy* **2015**, *73*, 36–45. [CrossRef]

83. Brigitte, Y.; Sauzedde, F.; Boillot, B.; Boddaert, S. Development of a building integrated solar photovoltaic thermal hybrid drying system. *Energy* **2017**, *128*, 755–767. [CrossRef]

84. Aristizábal, A.J.; Páez, C.A. Experimental investigation of the performance of 6 kW BIPV system applied in laboratory building. *Energy Build.* **2018**, *152*, 1–10. [CrossRef]

85. Chen, H.J.; Shu, C.M.; Chiang, C.M.; Lee, S.K. Energy Procedia The Indoor Thermal Research of the HCRI-BIPV Smart Window. *Energy Procedia* **2011**, *12*, 593–600. [CrossRef]

86. Wittkopf, S.; Valliappan, S.; Liu, L.; Seng, K.; Chye, S.; Cheng, J. Analytical performance monitoring of a 142.5 kWp grid-connected rooftop BIPV system in Singapore. *Renew. Energy* **2012**, *47*, 9–20. [CrossRef]

87. Macé, P.; Larsson, D.; Benson, J. *Transition towards Sound BIPV Business Models—Inventory on Existing Business Models, Opportunities and Issues for BIPV*; Technical Report; International Energy Agency—Photovoltaic Power Systems Programme: Paris, France, 2018.

88. Modules, P.; Walker, G.; Walker, G.R.; Sernia, P.C. Cascaded DC–DC Converter Connection of Photovoltaic Modules. *IEEE Trans. Power Electron.* **2004**, *19*, 1130–1139. [CrossRef]

89. Alonso, R.; Román, E.; Elorduizapatarietxe, S.; Ibáñez, P.; Apinaniz, S. Experimental Results of Intelligent PV Module for Grid-Connected PV Systems. In Proceedings of the 21st European Photovoltaic Solar Energy Conference and Exhibition, Dresden, Germany, 4–8 September 2006; Volume 53, p. 2297.

90. Walker, G.R.; Pierce, J.C. PhotoVoltaic DC-DC module integrated converter for novel cascaded and bypass grid connection topologies—Design and optimisation. In Proceedings of the PESC Record—IEEE Annual Power Electronics Specialists Conference, Jeju, Korea, 18–22 June 2006; pp. 1–7. [CrossRef]

91. Dong, M.; Dong, H.; Wang, L.; Yang, J.; Li, L.; Wang, Y. A Simple Open-Circuit Detection Strategy for a Single-Phase Grid-Connected PV Inverter Fed from Power Optimizers. *IEEE Trans. Power Electron.* **2018**, *33*, 2798–2802. [CrossRef]

92. Dong, H.; Yang, J.; Dong, M.; Su, M.; Cao, Y. Open-circuit fault detection for inverter fed by non-communication series-connected power optimizer. In Proceedings of the IECON 2017—43rd Annual Conference of the IEEE Industrial Electronics Society, Beijing, China, 29 October–1 November 2017; pp. 2606–2610. [CrossRef]

93. Ravyts, S.; Vecchia, M.D.; Zwysen, J.; Van den Broeck, G.; Driesen, J. Study on a cascaded DC-DC converter for use in building-integrated photovoltaics. In Proceedings of the 2018 IEEE Texas Power and Energy Conference (TPEC), College Station, TX, USA, 8–9 February 2018; pp. 1–6. [CrossRef]

94. Ravyts, S.; Vecchia, M.D.; Van den Broeck, G.; Driesen, J. Experimental Comparison of the Efficiency, Power Density and Thermal Performance of Two BIPV Converter Prototypes. In Proceedings of the 2018 IEEE 6th Workshop on Wide Bandgap Power Devices and Applications (WiPDA), Atlanta, GA, USA, 31 October–2 November 2018; pp. 7–13. [CrossRef]

95. Ravyts, S.; Dalla Vecchia, M.; Zwysen, J.; van den Broeck, G.; Driesen, J. Comparison Between an Interleaved Boost Converter Using Si MOSFETs Versus GaN HEMTs. In Proceedings of the PCIM Europe 2018; International Exhibition and Conference for Power Electronics, Intelligent Motion, Renewable Energy and Energy Management, Nuremberg, Germany, 5–7 June 2018; pp. 1–8.

96. Van De Sande, W.; Daenen, M.; Spiliotis, K.; Gonçalves, J.; Ravyts, S.; Saelens, D.; Driesen, J. Reliability Comparison of a DC-DC Converter Placed in Building-Integrated Photovoltaic Module Frames. In Proceedings of the 2018 7th International Conference on Renewable Energy Research and Applications (ICRERA), Paris, France, 14–17 October 2018; pp. 412–417. [CrossRef]

97. Engelen, K.; Shun, E.L.; Vermeyen, P.; Pardon, I.; D'hulst, R.; Driesen, J.; Belmans, R. The Feasibility of Small-Scale Residential DC Distribution Systems. In Proceedings of the 32nd Annual Conference on IEEE Industrial Electronics (IECON), Paris, France, 6–10 November 2006; pp. 2618–2623. [CrossRef]

98. Baran, M.E.; Mahajan, N.R. DC distribution for industrial systems: Opportunities and challenges. *IEEE Trans. Ind. Appl.* **2003**, *39*, 1596–1601. [CrossRef]

99. Agustoni, A.; Borioli, E.; Brenna, M.; Simioli, G.; Tironi, E.; Ubezio, G. LV DC distribution network with distributed energy resources: Analysis of possible structures. In Proceedings of the 18th International Conference and Exhibition on Electricity Distribution (CIRED), Turin, Italy, 6–9 June 2005; 5p. [CrossRef]

100. Justo, J.J.; Mwasilu, F.; Lee, J.; Jung, J.W. AC-microgrids versus DC-microgrids with distributed energy resources: A review. *Renew. Sustain. Energy Rev.* **2013**, *24*, 387–405. [CrossRef]

101. Elektrotechnik, D.K. *Deutsche Normungs-Roadmap Gleichstrom im Niederspannungsbereich Version 1*; Technical Report; Verband Der Elektrotechnik: Frankfurt, Germany, 2016; 138p.

102. Wang, H.; Blaabjerg, F. Reliability of capacitors for DC-link applications in power electronic converters—An overview. *IEEE Trans. Ind. Appl.* **2014**, *50*, 3569–3578. [CrossRef]

103. AlLee, G.; Tschudi, W. Edison Redux: 380 Vdc Brings Reliability and Efficiency to Sustainable Data Centers. *IEEE Power Energy Mag.* **2012**, *10*, 50–59. [CrossRef]

104. Rodriguez-Diaz, E.; Chen, F.; Vasquez, J.C.; Guerrero, J.M.; Burgos, R.; Boroyevich, D. Voltage-Level Selection of Future Two-Level LVdc Distribution Grids: A Compromise between Grid Compatibiliy, Safety, and Efficiency. *IEEE Electrif. Mag.* **2016**, *4*, 20–28. [CrossRef]

105. Chatterjee, U.; Gelagaev, R.; Masolin, A.; Driesen, J. Design of an intra-module DC-DC converter for PV application: Design considerations and prototype. In Proceedings of the IECON 2014—40th Annual Conference of the IEEE Industrial Electronics Society, Dallas, TX, USA, 29 October–1 November 2014; pp. 2017–2022. [CrossRef]

106. Michal, V. Optimal peak-efficiency control of the CMOS interleaved multi-phase step-down DC-DC Converter with segmented power stage. *IET Power Electron.* **2016**, *9*, 2223–2228. [CrossRef]

107. Mochizuki, M. Latest development and application of heat pipes for electronics and automotive. In Proceedings of the 2017 IEEE CPMT Symposium Japan (ICSJ), Kyoto, Japan, 20–22 November 2017; pp. 87–90. [CrossRef]

108. Flicker, J.; Tamizhmani, G.; Moorthy, M.K.; Thiagarajan, R.; Ayyanar, R. Accelerated testing of module-level power electronics for long-term reliability. *IEEE J. Photovolt.* **2017**, *7*, 259–267. [CrossRef]

109. Alam, M.K.; Khan, F.; Johnson, J.; Flicker, J. A Comprehensive Review of Catastrophic Faults in PV Arrays: Types, Detection, and Mitigation Techniques. *IEEE J. Photovolt.* **2015**, *5*, 982–997. [CrossRef]

110. Rezgui, W.; Mouss, H.; Mouss, N.; Mouss, D.; Benbouzid, M.; Amirat, Y. Photovoltaic module simultaneous open-and short-circuit faults modeling and detection using the I-V characteristic. In Proceedings of the IEEE International Symposium on Industrial Electronics, Buzios, Brazil, 3–5 June 2015; pp. 855–860. [CrossRef]

111. Kasper, M.; Ritz, M.; Bortis, D.; Kolar, J.W. PV Panel-Integrated High Step-up High Efficiency Isolated GaN DC-DC Boost Converter. In Proceedings of the Telecommunications Energy Conference 'Smart Power and Efficiency' (INTELEC), Hamburg, Germany, 13–17 October 2013; pp. 602–608.

112. Mamarelis, E.; Petrone, G.; Sahan, B.; Lempidis, G.; Spagnuolo, G.; Zacharias, P. What is the best dc/dc converter for an AC module? Experimental analysis of two interesting solutions. In Proceedings of the ISIE 2011: 2011 IEEE International Symposium on Industrial Electronics, Gdansk, Poland, 27–30 June 2011; pp. 1759–1764. [CrossRef]

Article

Evaluation of Supply–Demand Adaptation of Photovoltaic–Wind Hybrid Plants Integrated into an Urban Environment

Africa Lopez-Rey *, Severo Campinez-Romero, Rosario Gil-Ortego and Antonio Colmenar-Santos

Department of Electric, Electronic and Control Engineering, UNED, Juan del Rosal, 12-Ciudad Universitaria, 28040 Madrid, Spain; scampinez1@alumno.uned.es (S.C.-R.); rgil@ieec.uned.es (R.G.-O.); acolmenar@ieec.uned.es (A.C.-S.)
* Correspondence: alopez@ieec.uned.es; Tel.: +34-913-987-798

Received: 8 April 2019; Accepted: 6 May 2019; Published: 10 May 2019

Abstract: A massive integration of renewable energy sources is imperative to comply with the greenhouse emissions reduction targets fixed to achieve the limitation of global warming. Nevertheless, the present integration levels are still far from the targets. The main reason being the technical barriers arising from their non-manageable features. Photovoltaic and wind sources are the widest spread, as their maturity allows generation with a high-efficiency degree. A deep understanding of facilities' performance and how they can match the energy demand is mandatory to reduce costs and extend the technical limits and facilitate their penetration. In this paper, we present a novel methodology to evaluate how photovoltaic–wind hybrid facilities, placed in an urban environment can give generation patterns which will be able to match the demand profiles better than facilities installed individually. This methodology has been applied to a broad number of locations spread over the whole planet. The results show that with high homogeneity in terms of site weather characteristics, the hybrid facilities improve the matching up to 15% over photovoltaic plants and up to 35% over wind.

Keywords: wind energy; photovoltaic; complementarity; grid integration

---

1. Introduction

On 12 December 2015, the 195 countries participating in 21st Conference of the Parties (Paris Climate Change Conference) [1], organised by the United Nations Framework Convention on Climate Change (UNFCCC) [2], signed the Paris Agreement [3]. This agreement aims to achieve, as soon as possible, a reduction on the carbon emissions to hold the increase in the global average temperature to well below 2 °C above pre-industrial levels. The generation and use of energy are the main contributors to climate change, with 60% of the total greenhouse gases (GHG) emissions. The reduction in energy sector emissions is mandatory to achieve the global warming objectives. Hence, the Paris Agreement determines by 2030 there will be a substantial increase in the use of renewable energy sources (RES) in the world energy mix.

This important agreement is one more step given in the fight against climate change, which has been developed by the international community in the last decades. For this purpose, governments and international organisations and institutions have designed scenarios, strategies and commitments focused on the mitigation and reduction of the present emission levels. In all of them, high RES penetration shares are mandatory, and, with this aim, ambitious plans have been determined.

Along these lines, the United States of America developed the SunShot Initiative [4], focused on the solar photovoltaic renewable source (PV), favouring its integration by means of being competitive with the traditional generation forms before 2020, and the Wind Program [5], designed to speed up the development and integration of wind energy. Likewise, the member countries of the European Union established the Roadmap 2050 [6] to set up the paths to achieve the European commitment to reach in 2050 GHG emissions below 80% of 1990 levels.

Nowadays, the RES technologies with higher integration level are wind and photovoltaic. Figure 1 shows the time evolution of wind and PV installed power worldwide. At the end of 2017, the installed power capacity was 384 GW in PV facilities and 494 GW in wind farms.

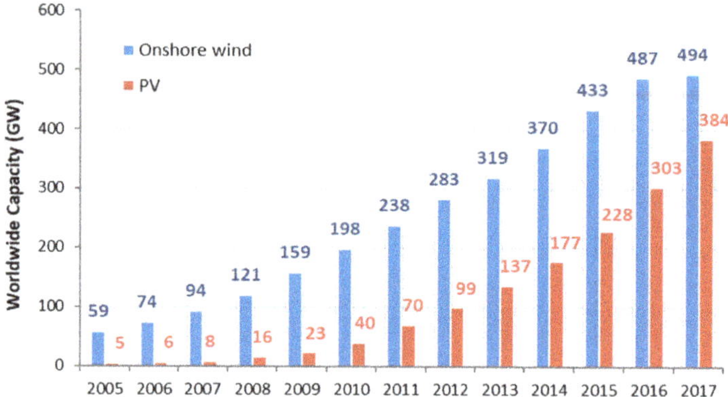

Figure 1. Wind and photovoltaic (PV) installed power worldwide. Source: [7,8].

Wind and PV electricity generation technologies presently offer technical and economic maturity levels. They allow high-efficiency generation almost everywhere at such a low cost compared with the traditional generation based on conventional thermal methods [9–15]. Moreover, among the renewable energy sources, wind and PV electricity generation technologies present high degrees of sustainability under multi-criteria analysis [16–18].

The International Energy Agency (IEA) remarks in its Energy Technology Perspectives 2017 (ETP2017) [19] that, the implementation of PV and onshore wind technologies are on-track to achieve their integration targets. Nevertheless, penetration shares for these technologies are still far from the targets fixed to contribute to the mitigation of GHG emissions. According to the IEA hi-Ren scenario (the high-renewables scenario—hi-Ren scenario—sees energy systems radically transformed to achieve the goal of limiting the global mean temperature increase to 2 °C target with a large share of renewables, which requires fast and strong deployment of photovoltaic and wind power and solar thermal electricity), the installed worldwide power capacity should reach 4674 GW by 2050 for PV and 2700 GW for onshore wind in the same period [20,21]. Innovative technical solutions and regulatory measurements are required to boost a massive RES integration to close the huge gap between the present status and the fixed targets in the next coming years.

The achievement of RES penetration targets is only feasible with actions addressed to facilitate their use in three fields with massive energy consumption: transport, buildings and industry. Among them, building integration shows the biggest potential to increase the share of RES in the energetic mix [22,23].

The widest field for RES building integration is found in the urban environment. Considerable research has been carried out to determine the PV [24–32] and wind potential [33–36] in urban areas and buildings.

PV presents a characteristic that favours its massive penetration in the urban environment: the dispersion degree. Solar radiation is received everywhere with such intensity levels that make possible the production of electricity. In addition, PV building integration offers environmental advantages as against its implementation on rural lands as the former gives a new value to the building roofs and facades.

In regard of wind energy, the installation of wind turbines in urban areas is not widely spread yet, but there are technical solutions to efficiently take advantage of the urban wind stream with its special characteristics of turbulence and direction variability [33,37–41].

In relation to PV–wind hybrid plants (PV+W hybrid hereinafter), extensive research has been developed to quantify the synergies between solar and wind sources. A non-exhaustive list of references is shown in Table 1.

Table 1. Literature review reference list.

Topic	Reference
Smoothing resource and the correlation between the wind and solar PV resource	[42]
Variability and determination of regional or local wind solar complementarity or synergy	[43–47]
Determination of flexibility requirements of large-scale wind and PV penetration	[48]
Impact of wind solar complementarities on storage sizing and use	[49]
Effect of solar and wind resources complementarity in micro-hybrid system reliability	[50]

Cities are big electricity consumers. Therefore, RES integration in urban areas would also offer an important technical advantage because the generation would be placed near to the consumption point. This solution would improve the whole electric system efficiency by reducing the transport and distribution of electricity losses. Moreover, it is a clear example of distributed generation with advantages associated with the control and management of the electric network [51–53].

But the integration of a massive share of variable RES (VRES) in the electric power grid implies technical challenges and extra-costs. The electricity generated in PV and wind facilities have a non-manageable character; which means that it is not possible to control the supply instantaneously (except to reduce it) to match the demand. A high VRES penetration requires the application of measures focused on planning, operation and flexibility of the whole system to respond to the uncertainty and variability in the supply–demand balance in short timescales [54–56]. These measures present estimable costs for the system that could reach 25–35 €/MWh in high penetration scenarios [57,58].

Extensive research has been recently carried out showing that, with the use of adequate coordination control algorithms, large-scale systems made up of multiple individual subsystems can together contribute efficiently in the achievement of global quantities of interest, even in the case that some of the sub-systems became adversarial or non-cooperative due to bad functioning [59]. This resilient performance is fully applicable to a massive integration of VRES based on the implementation of individual small facilities.

Due to the aforementioned, a deep knowledge of the performance of the facilities and their generation patterns becomes relevant. It is essential to understand how they could match the electricity demand, with the aim to offer better control and management of the electricity fed into the grid and, consequently, collaborate to reduce the technical barriers and to decrease the integration cost.

With this target as the main objective of our work, we have carried out a study under the novel perspective to evaluate the supply–demand balance adaptation of PV+W hybrid plants integrated into an urban environment. To have results applicable on a global scale, we have considered hundreds of locations spread all over the world and multiple load profiles for the characterisation of demand. This article first analyses if PV+W hybrid facilities present generation patterns that adapt better to the demand profiles than if the facilities were installed individually, and second, determines a novel methodology to quantify the adaptation degree.

The novelty of our work is fundamentally based on three main grounds:

- The evaluation of supply–demand balance adaptation of PV+W hybrid plants
- The hybrid plants are integrated into an urban environment
- The results are applicable on a global scale as we have considered real weather data from hundreds of locations spread all over the world and multiple profiles for the characterisation of the demand.

The main technical challenge arises from our requirement to obtain results applicable on a global scale. With that aim, we have considered only real weather data from hundreds of meteorological stations and multiple electricity load profiles for the characterisation of the demand in different seasons and days. These requirements have obliged the authors to carry out extensive work to obtain and validate the input data and get it homogeneous.

Below in Section 2, we introduce the methodology developed to evaluate and quantify the level of adaptation of generation patterns to demand profiles. In Section 3, we present the results of applying this methodology to a wide number of locations worldwide and carry out a sensitivity analysis of the results. Finally, in Section 4, the conclusions of our study are discussed.

2. Methodology

Our work aims to analyse if the generation patterns of PV+W hybrid facilities match better with the demand profiles than if the facilities were considered separately. We will not determine what would be the absolute coverage of electricity that the facilities could provide to the whole electric demand. This approach is like evaluating to what extent the generation and demand curves have the same "shape".

We propose the evaluation of the adaptation level by the determination of the matching factor (ε). It will be calculated as the average quadratic error between the electric generation patterns and the demand profiles, previously normalised and particularised for every single location under study, as will be detailed below. In this way, ε would be zero when the adjustment is perfect; it means, when the generation and demand curves have the same shape, and ε would rise to one when the difference becomes higher. The proposed methodology to calculate ε is illustrated in Figure 2.

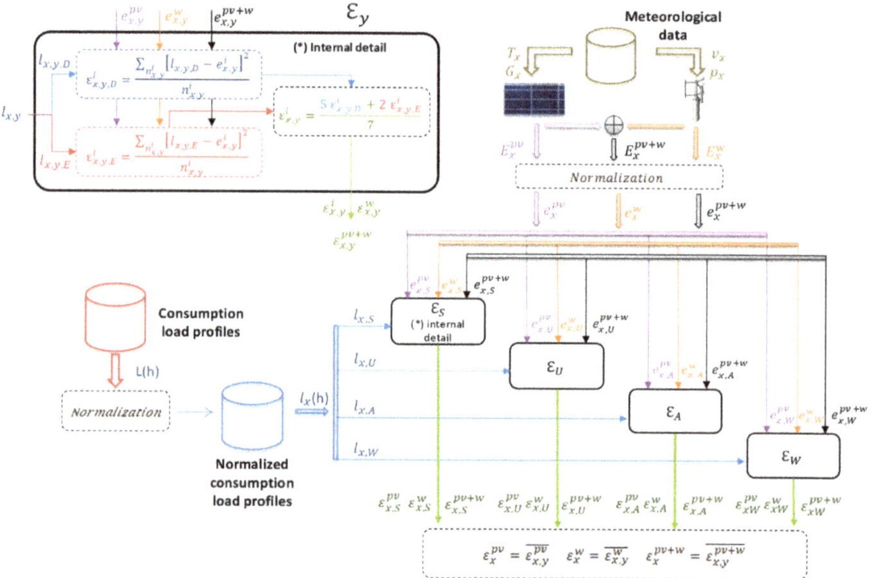

Figure 2. Methodology for obtaining the matching factor (ε) at a single location.

The calculation starts with the collection of hourly climate data representative of an average year in every single location included in the analysis. The data collected have been pressure p, temperature T, wind speed v and irradiation G. To generalise the results, it is essential to count on climatologic data for multiple locations spread around the Earth. With this data, together with the dimensioning and characterisation of a PV+W hybrid facility, we obtain yearly patterns of the foreseen generation for every type of facility E_x^{pv}, E_x^{p} y E_x^{pv+w}.

Second, electricity demand profiles are required. To select the appropriate load profiles to be utilised in the calculation of the matching factor, it is required to quantify to what extent the hybrid facilities' generation would contribute to the country-level aggregated load. With this aim, first we have done a rough estimation of the amount of electricity that could be generated by hybrid facilities placed in an urban environment (buildings) in a scenario of high penetration, and, second, we have calculated the aggregated demand coverage on hourly basis. The hourly aggregated demand coverage was calculated by using Equation (1):

$$Hourly\ aggregated\ demand\ coverage = \frac{NB \times AB \times CF \times AHIC}{RHAD} \tag{1}$$

where:

- NB is the number of buildings in the relevant country or region
- AB is the share of available buildings in the relevant country or region, defined as those buildings where the installation of a PV+W hybrid facility would be feasible.
- CF is the capacity factor of the PV+W hybrid facility, defined for one specific period as the electricity generated by the hybrid facility in one hour divided into its installed capacity.
- $AHIC$ is an average PV+W hybrid installed capacity.
- $RHAD$ is the hourly aggregated demand representative for the country or area under analysis.

To have an estimate in different scenarios, the calculation of the hourly aggregated demand coverage was done for two regions (Europe and the United States) and a European country (Spain) Table 2 shows the specifics of each region or country considered in the calculations.

Table 2. Region and country specifics for hourly aggregated demand coverage calculation.

Variable	Spain (SP)	Europe EU28 Countries (EU)	US
NB	10,000,000 [60,61]	130,000,000 [62]	142,500,000 [63,64]
CF		50%	
RHAD (MWh)	30,000 [65]	400,000 [66,67]	430,000 [68]

Table 3 shows the hourly aggregate demand coverage of the hybrid facilities for different values of (i) $AHIC$ and (ii) AB. To obtain conservative values, it was set up 50% of CF and limits of 15% for AB and 10 kW for the $AHIC$. The results show that shares around 10% of hourly aggregated demand coverage could be reached with moderated values of AB and $AHIC$. The hourly coverage might reach levels over 20% in more optimistic scenarios.

Table 3. Hourly aggregated demand coverage.

		AHIC (kW)										
		2			5			7.5			10		
Country/Region		SP	EU	US	SP	EU	US	SP	EU	US	SP	EU	US
	5.0%	1.7%	1.6%	1.7%	4.2%	4.1%	4.1%	6.3%	6.1%	6.2%	8.3%	8.1%	8.3%
AB (%	7.5%	2.5%	2.4%	2.5%	6.3%	6.1%	6.2%	9.4%	9.1%	9.3%	12.5%	12.2%	12.4%
share out	10.0%	3.3%	3.3%	3.3%	8.3%	8.1%	8.3%	12.5%	12.2%	12.4%	16.7%	16.3%	16.6%
of total)	12.5%	4.2%	4.1%	4.1%	10.4%	10.2%	10.4%	15.6%	15.2%	15.5%	20.8%	20.3%	20.7%
	15.0%	5.0%	4.9%	5.0%	12.5%	12.2%	12.4%	18.8%	18.3%	18.6%	25.0%	24.4%	24.9%

The level of coverage obtained should be considered in the management of ancillary services and market operations. Based on the above, aggregated load profiles have been selected in the calculation of the matching factor.

The demand evolution presents a high dependency on the climate, the distribution of the working days and the consumer´s habits. One of the objectives of this study is to obtain results applicable globally. Hence, we have utilised multiple profiles to characterise the electricity consumption everywhere. The methodology here proposed includes the determination of 16 different hourly demand curve profiles, as shown in Table 4, distinguishing between (i) the Northern or Southern hemisphere, (ii) the year season and (iii) weekdays and weekends (bank holidays are included in the weekend day category). Based on the above, the demand profiles used in the calculation for every location will be the eight corresponding to the hemisphere where the location is placed.

Table 4. Hourly demand profiles.

Hemisphere	Day	Season			
		Spring (S)	Summer (U)	Autumn (A)	Winter (W)
Northern (N)	Weekday (D)	L_{NSD}	L_{NUD}	L_{NAD}	L_{NWD}
	Weekend (E)	L_{NSE}	L_{NUE}	L_{NAE}	L_{NWE}
Southern (S)	Weekday (D)	L_{SSD}	L_{SUD}	L_{SAD}	L_{SWD}
	Weekend (E)	L_{SSE}	L_{SUE}	L_{SAE}	L_{SWE}

As has been discussed before, our methodology is applied to quantify the adaptation degree of the generation to the aggregate demand (i.e., for a country) and not only to local demand where the facility is placed (household, garage, shopping centre, etc.). However, the absolute generation level of every facility, even the aggregation of a high number of them cannot be compared to the global, regional or national demand. We are, therefore, obliged to include in the methodology a mechanism to eliminate the scale effect from ε calculation. The way we propose here is to determine normalised patterns for both generation and demand profiles as follows:

1. Both demand profiles and generation patterns are considered on an hourly basis.
2. The normalisation period for generation and demand is daily.
3. The normalised demand profiles are obtained by dividing each hourly data into the respective daily maximum.
4. The individual normalised PV and wind daily generation profiles are obtained by dividing each hourly data into the respective daily maximum.
5. Three normalised generation profiles for the hybrid facility are obtained as per the following methods:

- Method 1: By adding the individual PV and W (wind) normalised profiles:

$$e_x^{pv+w} = e_x^{pv} + e_x^{w} \tag{2}$$

- Method 2: By dividing every hourly data into the maximum value of both facilities.

$$e_x^{pv+w} = \frac{E_x^{pv+w}}{max\left(E_x^{pv}, E_x^{w}\right)} \tag{3}$$

- Method 3: By dividing every hourly data into the daily maximum value of the hybrid facility.

$$e_x^{pv+w} = \frac{E_x^{pv+w}}{max\left(E_x^{pv+w}\right)} \tag{4}$$

These three methods to normalise the values of the hourly hybrid generation profiles do not pretend to have a physical sense by themselves. Our methodology is oriented to find out how the matching factor ε changes when the PV and wind facilities are considered together in a hybrid plant. With this aim, what is relevant to quantify this change is to evaluate it by using the results obtained with the same normalisation method.

Figure 3 shows, as an example, the normalised curves for one day in the period under analysis, where it can be seen:

- The normalised demand hourly profiles for a weekday $l_{x,y,D}$ and for a weekend day $l_{x,y,E}$.
- The normalised generation hourly patterns for the PV facility e_x^{pv} and the wind one e_x^{w}.
- Three hourly generation patterns of the hybrid facility e_x^{pv+w}, each one normalised according to the corresponding method.

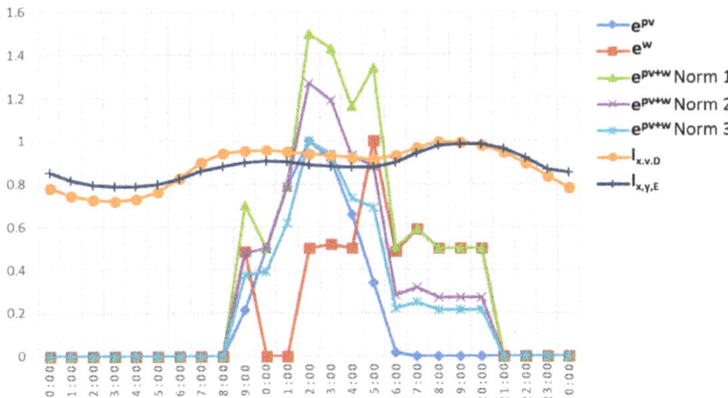

Figure 3. One-day-example of the normalised generation and load profile evolution.

Once the normalised hourly patterns are determined, ε is calculated for every single location by following the next steps:

1. The relevant eight normalised demand profiles are selected according to the site location in the Northern or the Southern hemisphere (Table 4).
2. For every annual season, ε is calculated for weekdays (5) and for weekend days (6). The weighted average value is calculated using (7):

$$\varepsilon_{x,y,D}^i = \frac{\sum n_{x,y}^i \left[l_{x,y,D} - e_{x,y}^i \right]^2}{n_{x,y}^i} \tag{5}$$

$$\varepsilon_{x,y,E}^i = \frac{\sum n_{x,y}^i \left[l_{x,y,E} - e_{x,y}^i \right]^2}{n_{x,y}^i} \tag{6}$$

$$\varepsilon_{x,y}^i = \frac{5\,\varepsilon_{x,y,D}^i + 2\,\varepsilon_{x,y,E}^i}{7} \tag{7}$$

where:

- $\varepsilon_{x,y,D}^i$ and $\varepsilon_{x,y,E}^i$ are the matching factors in weekdays D and weekend days E, respectively, for the facility type i, placed at the location x, during the season y.
- $n_{x,y}^i$ is the number of hours in the season y at the location x.

- $l_{x,y,D}$ and $l_{x,y,E}$ are the normalised demand profiles in weekdays D and weekend days E, respectively, at the hemisphere where is placed the location x, during the season y.
- $e^i_{x,y}$ is the generation pattern for the facility type i, placed at the location x, during the season y.
- $\varepsilon^i_{x,y}$ is the matching factor of the generation facility type i, placed at the location x, during the season y.

3. Finally, the yearly matching factor for each type of facility and location is obtained by averaging the factors calculated for every season as per (8):

$$\varepsilon^i_x = \overline{\varepsilon^i_{xy}} = \frac{\varepsilon^i_{x,S} + \varepsilon^i_{x,U} + \varepsilon^i_{x,A} + \varepsilon^i_{x,W}}{4} \tag{8}$$

2.1. Climatic Data

The climate raw data used in this article has been obtained from the Meteonorm database [69]. This commercial software provides, for an average climatic year, among other variables: hourly data of pressure, temperature, superficial wind speed and solar irradiation incident on an optimally tilted solar panel.

Meteonorm provides weather data everywhere on the planet by means of the interpolation of registered variables in specific points. However, we have only used those locations where the meteorological stations are placed and are logging the climatic variables directly. With this criterion, 844 locations spread over the whole planet were selected.

With the objective to generalise the results of the application of the methodology, the selected locations have been classified following the Köppen–Geiger climatic regions, which divides the Earth into regions according to their weather conditions [70,71]. Figure 4 shows the location of the meteorological stations used in this study and their correspondence with the Köppen–Geiger regions.

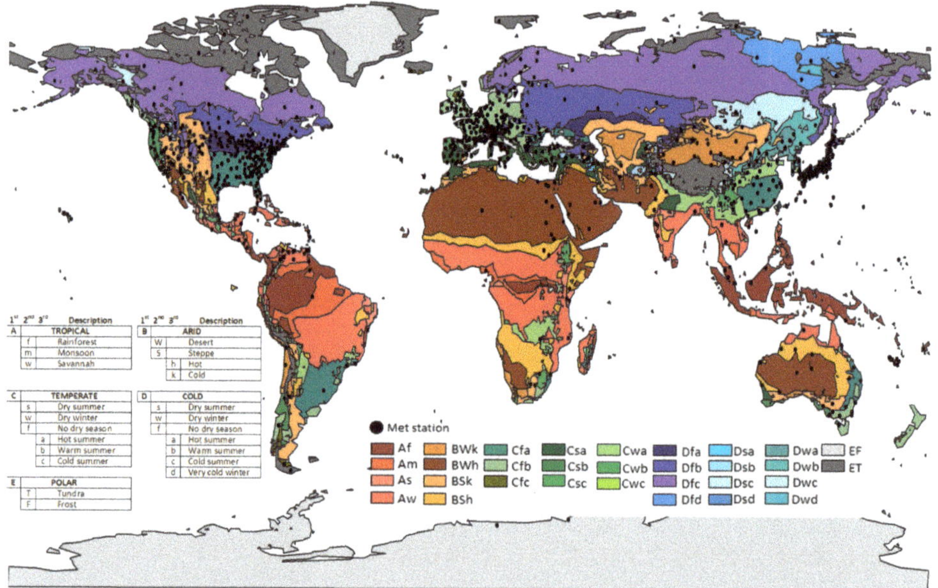

Figure 4. Location of the meteorological stations in the Köppen–Geiger climate classification areas. Source: [72] and self-elaboration.

2.2. Solar PV and Wind Generation Patterns

To estimate the electricity generation, a PV+W hybrid facility prototype has been designed according to the simplified diagram shown in Figure 5.

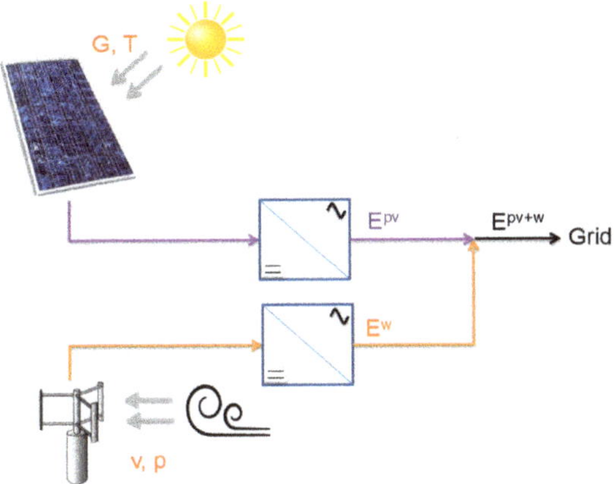

Figure 5. Single line diagram of the photovoltaic+wind hybrid (PV+W hybrid) facility.

The electricity produced by the PV facility placed at the location x is calculated with the following expression, adapted from [73]:

$$E_x^{pv} = G_x \cdot A_{PV} \cdot PR \cdot \eta \cdot [1 + \alpha(T_x - 293)]$$ (9)

where:

- G_x is the total solar irradiation incident on an optimally tilted solar panel.
- A_{PV} the surface covered by solar panels.
- η is the solar PV panel efficiency.
- PR is the facility performance ratio.
- α is the maximum power temperature coefficient.
- T_x is the ambient temperature (the temperature coefficient should be applied to the difference between the solar panel temperature and the standard value of 293 K. Nevertheless, as the solar temperature is not available, the correction has been applied considering the ambient temperature).

Nowadays there are different technologies used in the manufacturing of solar panels; the most widely used is multi-crystalline silicon cells [74]. For the calculation of the electricity generation, it was selected a commercial solar panel manufactured with multi-crystalline silicon cells and an efficiency η of 15.5%. The rest of the solar panel characteristics are shown in Table 5.

Table 5. Solar panel characteristics [75].

Characteristic	Value
Manufacturer	Trina Solar
Model	TSM-PC14
Cell type	Si Multicrystalline
Maximum Power (STC conditions)	300 W
Efficiency (η)	15.5%
Dimensions (h × w × d)	1956 × 992 × 40 mm^3
Temperature Coefficient of maximum power (α)	−0.41%/K

The PV facilities present current *PR* values in the 60 to 90% range [76,77], therefore, in this study, a mean value of 75% was considered for *PR*.

The area for solar panels was set up in 23.2 m^2 because it is a medium size surface suitable to be placed on every roof, pergola, etc. According to the characteristics of the solar panel selected, this area means 12 solar panels giving a power capacity of 3.6 kW.

For the wind facility, a vertical-axis wind turbine generator (VAWT) was selected. These types of wind turbines are more efficient in locations where the wind stream presents both high turbulence and continuous variations in the direction, such as in the urban environment [33,37,41]. The VAWT considered in the calculations has a nameplate power of 3.5 kW, similar to the PV installed capacity. Figure 6 shows the VAWT power curve for standard density (ρ_{std} = 1.225 kg/cm^2).

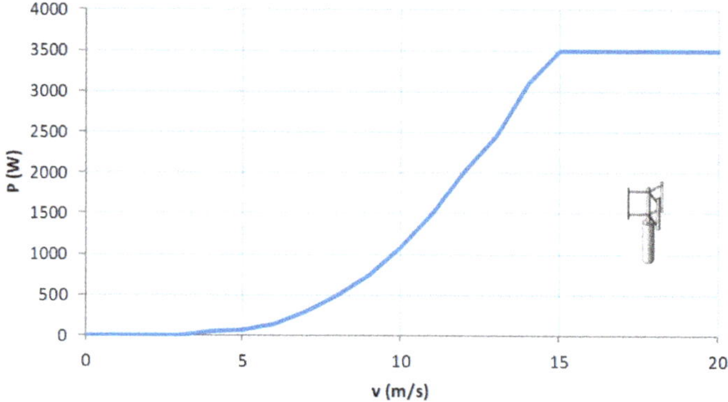

Figure 6. Vertical-axis wind turbine generator (VAWT) power curve for the standard air density ρ_{std} = 1.225 kg/cm^2. Source [78].

The electricity produced by the wind facility is calculated according to the following equation (adapted from [79]):

$$E_x^w = \rho \cdot \frac{P_x}{\rho_{std}} \cdot t \tag{10}$$

where:

- *P* is the output power from the power curve corresponding with the wind speed incident on the el VAWT (Figure 6).
- ρ is the air density.
- *t* is the time.

Finally, the electricity produced by the PV+W hybrid facility is:

$$E_x^{pv+w} = E_x^{pv} + E_x^w \tag{11}$$

The energy produced was calculated for each type of facility (PV, wind and PV+W hybrid) in all locations, obtaining the evolution in an average year with climatic conditions characterised for the variables defined in Chapter 2.1. Figure 7 shows, as an example, the generation curves of the PV, wind and hybrid facilities in an average month of May at one of the locations considered in this study.

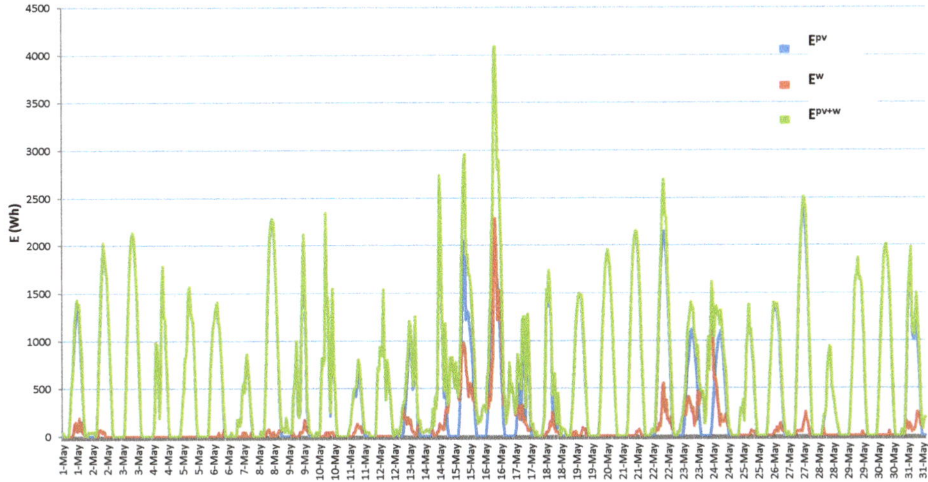

Figure 7. Time evolution of the generation in May.

One interesting result of this first step of the calculation is the contribution of the electricity sources, PV and wind, to the total hybrid facility production. Table 6 shows the PV Facility contribution to the total generation in the group of different climatic regions. Despite the PV and wind capacity being similar, the contribution of PV is a majority with 84% on average; going from 71% in polar climate zones to 91% in tropical areas. This predominance of PV is justified because the facility locations were not chosen with the criterion of having a relevant wind resource.

Table 6. PV contribution to the PV+W hybrid-facility generation. (See Figure 4 for climate zone codification).

Climate Zone	n° Stations	$\frac{E_x^{pv}}{E_x^{pv+w}}$	Climate Zone	n° Stations	$\frac{E_x^{pv}}{E_x^{pv+w}}$
Arid	109	0.89	*Tropical*	66	0.91
BSh	16	0.90	Af	21	0.89
BSk	46	0.86	Am	8	0.93
BWh	27	0.90	As	4	0.87
BWk	20	0.92	Aw	33	0.92
Cold	188	0.83	*Temperate*	459	0.84
Dfa	29	0.78	Cfa	192	0.87
Dfb	90	0.83	Cfb	162	0.78
Dfc	40	0.81	Cfc	4	0.68
Dfd	4	0.93	Csa	45	0.85
Dsa	1	1.00	Csb	34	0.87
Dsb	3	0.93	Cwa	17	0.91
Dsc	1	0.82	Cwb	5	0.96
Dwa	10	0.88			
Dwb	5	0.88	*Polar*	32	0.71
Dwc	5	0.95	EF	2	0.33
			ET	30	0.74
Global				854	0.84842

2.3. Demand Load Profiles

The demand profiles were defined using real data provided from the commercial companies and distributor and transport system operators detailed in Table 7.

Table 7. Demand load profiles data source [65,80–84].

Hemisphere	Distributor/Operator	Country
North	Red Eléctrica de España	Spain
	PJM	USA–Northeast
	Midcontinent Independent System Operator	USA–West
	Northwest PowerPool	USA–Northwest
South	National Electricity Coordinator	Chile
	Australian Energy Market Operation	Western Australia

From all the sources, real hourly demand curves for the 365 days of 2015 were obtained. Then, to determine the sixteen standard demand profiles used in ε calculation (Table 4), the next steps were followed:

1. The curves from every load profile were normalised dividing each hourly data into its respective daily maximum.
2. Once normalised, the curves were separated out from the season and from weekday and weekend days.
3. It was obtained average normalised curves for both hemispheres.

The normalised demand profiles obtained are shown in Figures 8–11.

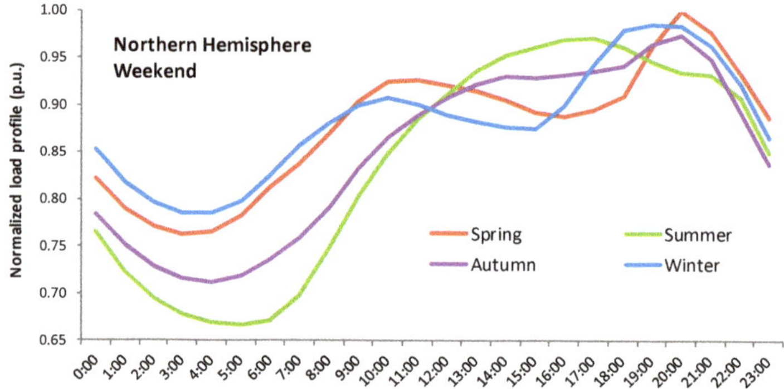

Figure 8. Standard normalised demand profiles for weekend days in the Northern Hemisphere. Source: [65,80–82] and self-elaboration.

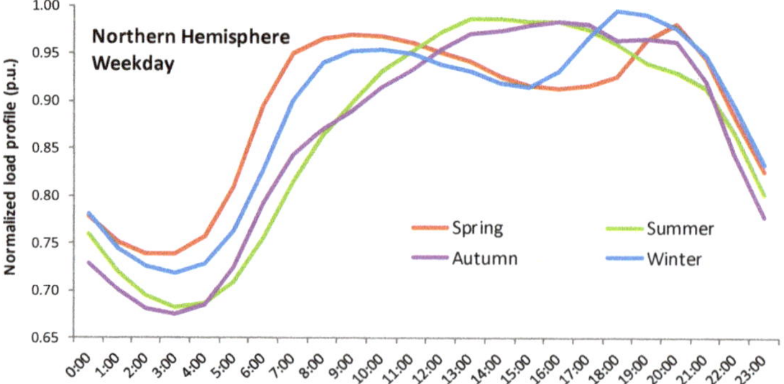

Figure 9. Standard normalised demand profiles for weekdays in the Northern Hemisphere. Source: [65,80–82] and self-elaboration.

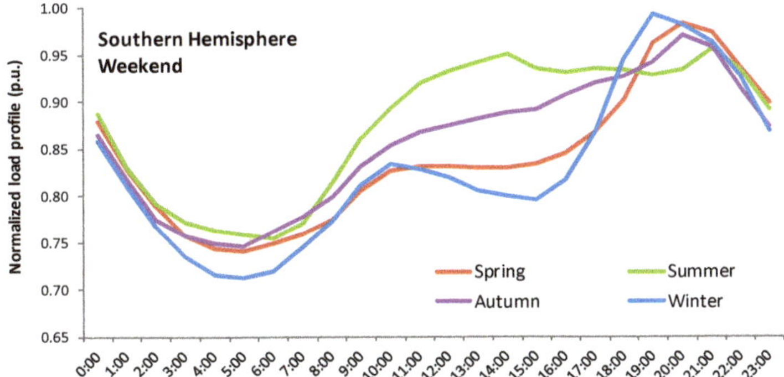

Figure 10. Standard normalised demand profiles for weekend days in the Southern Hemisphere. Source: [83,84] and self-elaboration.

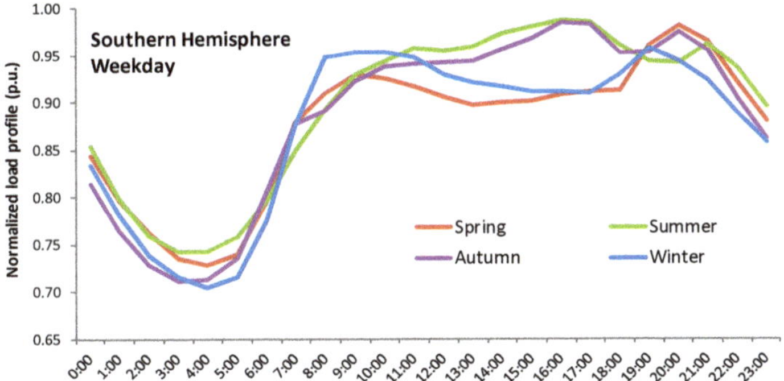

Figure 11. Standard normalised demand profiles for weekdays in the Southern Hemisphere. Source: [83,84] and self-elaboration.

3. Results

Once the normalised generation patterns and demand profiles have been determined, it is possible to obtain ε by applying Equations (5)–(8). The calculation was made individually for the 844 locations defined in Section 2.1 by using a Microsoft VBA macro programme in Excel.

The results, sorted by the Köppen–Geiger climate areas, are shown in Table 8.

The global matching factor obtained for PV facilities ε^{PV} is 0.46. As it can be noted, this value is quite homogeneous in all the climatic regions.

The global matching factor for the wind facilities ε^{W} is 0.6, that means 30% worse adaptation to demand profiles than PV plants. The results present a low dispersion degree with respect to the climatic areas. The minimum value of 0.56 is obtained for polar climates (-5% out of global value), while the maximum, 0.63, is found for tropical climates ($+7\%$ out of global).

For PV+W hybrid plants, depending on the normalisation method, the results obtained for ε^{PV+W} go from 0.4 if the method 1 is used, to 0.42 if the method 2 is used and 0.43 if the method 3 is used. Once again, the minimum factor is obtained for sites located in polar climates and the maximum for tropical areas. The degree of dispersion is also very negligible.

Figure 12 illustrates the comparison of the matching factor for the PV+W hybrid plants ε^{PV+W} versus PV facilities ε^{PV}. As it can be noted, in a global context, the adaptation of the hybrid facility is 15% higher for the method 1, 9% higher for the method 2 and 7.7% for the method 3. The highest improvement is given for polar climate areas and the lowest for arid and tropical areas.

Table 8. ε^{pv}, ε^{w} and ε^{pv+w} for every single individual Köeppen–Geiger climatic regions (See Figure 4 for climate zone codification).

Climate Zone	n° Stations	ε^{pv}	ε^{w}	Method 1 $e^{pv+w}=e^{pv}+e^{w}$			Method 2 $e^{pv+w}=\dfrac{E^{pv+w}}{max(E^{pv},E^{w})}$			Method 3 $e^{pv+w}=\dfrac{E^{pv+w}}{max(E^{pv+w})}$		
				ε^{pv+w}	$\dfrac{\varepsilon^{pv+w}}{\varepsilon^{pv}}$	$\dfrac{\varepsilon^{pv+w}}{\varepsilon^{w}}$	ε^{pv+w}	$\dfrac{\varepsilon^{pv+w}}{\varepsilon^{pv}}$	$\dfrac{\varepsilon^{pv+w}}{\varepsilon^{w}}$	ε^{pv+w}	$\dfrac{\varepsilon^{pv+w}}{\varepsilon^{pv}}$	$\dfrac{\varepsilon^{pv+w}}{\varepsilon^{w}}$
Arid	**109**	**0.46**	**0.60**	0.40	−12%	−33%	0.43	−6%	−28%	0.43	−5%	−27%
BSh	16	0.46	0.59	0.40	−12%	−32%	0.43	−6%	−27%	0.43	−5%	−26%
BSk	46	0.46	0.59	0.39	−14%	−33%	0.42	−8%	−28%	0.43	−6%	−27%
BWh	27	0.45	0.58	0.40	−12%	−31%	0.43	−6%	−26%	0.43	−5%	−25%
BWk	20	0.45	0.65	0.42	−6%	−35%	0.44	−3%	−32%	0.44	−3%	−32%
Cold	**188**	**0.47**	**0.61**	0.40	−6%	−35%	0.42	−9%	−30%	0.43	−8%	−29%
Dfa	29	0.47	0.55	0.37	−20%	−33%	0.41	−12%	−26%	0.42	−11%	−25%
Dfb	90	0.47	0.60	0.39	−16%	−35%	0.42	−10%	−30%	0.43	−8%	−29%
Dfc	40	0.47	0.62	0.40	−16%	−36%	0.42	−11%	−32%	0.43	−10%	−31%
Dfd	4	0.47	0.71	0.46	−4%	−37%	0.47	−2%	−35%	0.47	−1%	−35%
Dsa	1	0.46	0.76	0.46	0%	−40%	0.46	0%	−40%	0.46	0%	−40%
Dsb	3	0.46	0.65	0.42	−9%	−36%	0.44	−4%	−32%	0.45	−3%	−32%
Dsc	1	0.47	0.60	0.37	−20%	−38%	0.41	−13%	−32%	0.41	−11%	−31%
Dwa	10	0.47	0.64	0.42	−10%	−35%	0.44	−6%	−31%	0.44	−5%	−31%
Dwb	5	0.46	0.64	0.41	−11%	-36%	0.44	−6%	−32%	0.44	−5%	−31%
Dwc	5	0.45	0.68	0.43	−5%	−37%	0.44	−2%	−35%	0.45	−2%	−35%
Polar	**32**	**0.47**	**0.56**	0.37	−22%	−34%	0.38	−19%	−31%	0.39	−17%	−30%
EF	2	0.49	0.34	0.26	−48%	−24%	0.25	−49%	−27%	0.26	−48%	−25%
ET	30	0.47	0.57	0.37	−20%	−34%	0.39	−16%	−31%	0.40	−15%	−30%
Temperate	**459**	**0.47**	**0.61**	0.39	−15%	−35%	0.42	−9%	−30%	0.43	−8%	−29%
Cfa	192	0.47	0.60	0.40	−15%	−34%	0.43	−8%	−29%	0.43	−7%	−28%
Cfb	162	0.47	0.59	0.38	−19%	−36%	0.41	−13%	−31%	0.42	−11%	−29%
Cfc	4	0.47	0.54	0.35	−25%	−35%	0.39	−18%	−29%	0.39	−16%	−28%
Csa	45	0.46	0.62	0.41	−11%	−35%	0.43	−7%	−31%	0.43	−6%	−30%
Csb	34	0.46	0.63	0.41	−12%	−36%	0.43	−6%	−32%	0.44	−5%	−31%
Cwa	17	0.46	0.64	0.42	−10%	−35%	0.44	−5%	−31%	0.44	−4%	−30%
Cwb	5	0.45	0.67	0.43	−5%	−36%	0.44	−2%	−34%	0.45	−2%	−33%
Tropical	**66**	**0.46**	**0.63**	0.41	−10%	−34%	0.44	−5%	−30%	0.44	−4%	−29%
Af	21	0.46	0.62	0.41	−11%	−34%	0.43	−6%	−29%	0.44	−5%	−29%
Am	8	0.46	0.64	0.42	−9%	−35%	0.44	−4%	−31%	0.45	−3%	−30%
As	4	0.46	0.55	0.39	−16%	−30%	0.42	−8%	−23%	0.43	−6%	−22%
Aw	33	0.46	0.64	0.42	−9%	−35%	0.44	−5%	−31%	0.44	−4%	−30%
Global	**854**	**0.46**	**0.60**	0.4	−15%	−35%	0.42	−8.9%	−30%	0.43	−7.7%	−29%

The comparison of the matching factor for the PV+W hybrid facility ε^{PV+W} versus wind ε^{W} is shown in Figure 13. The adaptation is much higher in this case than when it is compared with the PV facility; as it has obtained an improvement of 35% for the method 1, 30% for the method 2 and 29% for the method 3. The values are quite similar in all the climate areas.

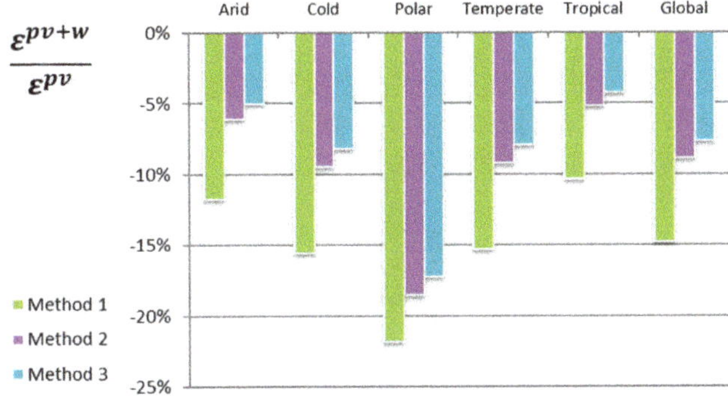

Figure 12. Improvement of the matching factor: ε^{pv+w} (hybrid) over ε^{pv} (solar PV) facilities.

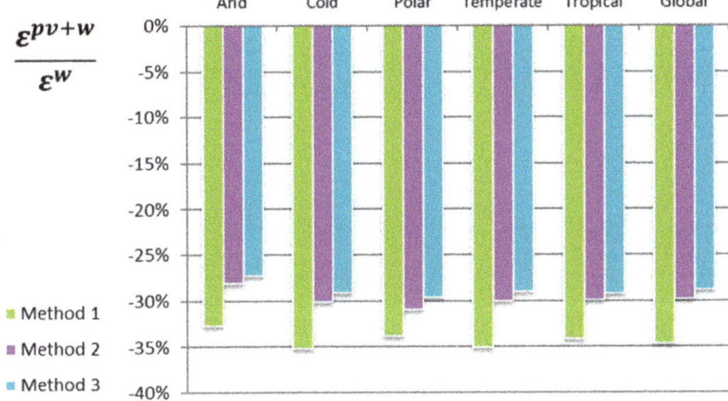

Figure 13. Improvement of the matching factor: ε^{pv+w} (hybrid) over ε^{w} (wind) facilities.

3.1. Sensitivity Analysis

It is mandatory to check if the methodology here proposed would give stable results in case of the variation of the relevant variables considered in the calculations. The technical characteristics of the facilities, as well as the performance parameters of the equipment, are quite steady and will be under control with adequate maintenance. The more relevant variations can arise from (i) deviations or errors in the evaluation of the solar and wind resource at the location or the use of non-optimised facilities (i.e., tilt or azimuth angles of the PV facility different from the ideal) and (ii) different power capacity of the facilities. In this way, to determine the robustness of the methodology two sensitivity analyses were carried out with respect to those variables.

3.1.1. Sensitivity Related to Errors in the Resource Valuation

To evaluate the variations in the valuation of the resource produced by errors, spatial smoothing effector the installation of the facilities (non-optimisation), the electricity generated is calculated by means of a modification of the Formulas (9) and (10) to include the multiplying factors f_{pv} and f_w to simulate the variation of the solar irradiance and wind resource. The methodology was applied for a wide variation range of the multiplying factors between 0.7 to 1.3 which represents a variation of $\pm30\%$ in the renewable resources.

$$E_x^{pv}\left(f_{pv}\right) = f_{pv}{\cdot}G_x{\cdot}A_{PV}{\cdot}PR{\cdot}\eta{\cdot}[1 + \alpha(T_x - 293)] \tag{12}$$

$$E_x^w(f_w) = \rho \cdot \frac{P_x(f_w)}{\rho_{std}} \cdot t \qquad (13)$$

The variation of ε^{PV+W} with the multiplication factors is illustrated in the Figure 14. As it can be shown, the methodology is robust because:

1. ε^{PV+W} hardly varies with changes of the irradiation for the three normalisation methods.
2. The effect of variations in wind resource is quite limited. For increases in the mean wind speed of 30% ($f_w = 1.3$) ε^{PV+W} rises about 5%, while a decrement of 30% ($f_w = 0.7$) produces a variation range from −5%, (normalisation method 3) to −8% (normalisation method 1).

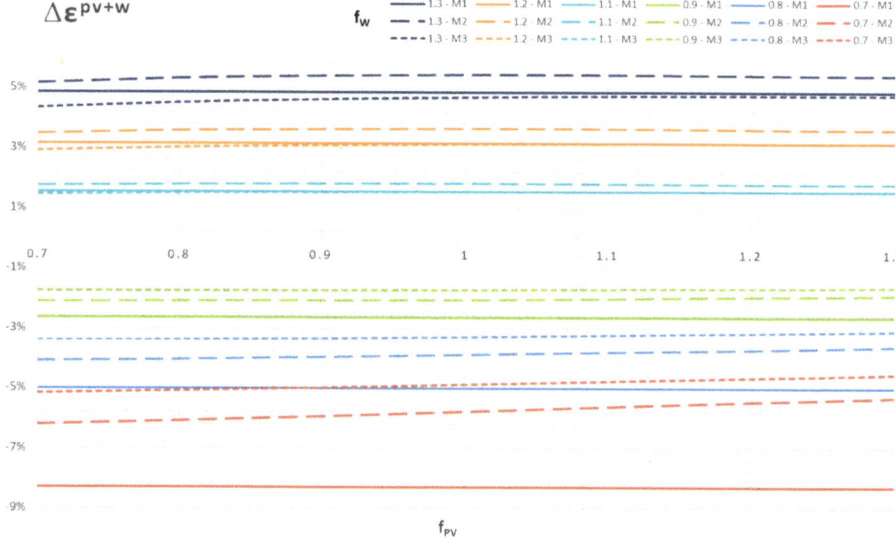

Figure 14. Variation of ε^{PV+W} with the multiplication factors f_{pv} and f_w. (M1, M2 and M3 represent the results obtained by means of application of the normalisation methods 1, 2 and 3, respectively).

3.1.2. Sensitivity Related to the Power Capacity of the Facilities

We applied the methodology considering generation patterns of a PV+W hybrid facility with twice the power capacity of the facility previously considered to evaluate the potential variations in the results produced by changes on the installed power capacity of the wind and PV facilities. The solar panel and VAWT used now have the following characteristics:

1. A commercial solar panel manufactured with multi-crystalline silicon cells and an efficiency η of 17.5%. The rest of the solar panel characteristics are shown in the Table 9.

Table 9. Solar panel characteristics [85].

Characteristic	Value
Manufacturer	Trina Solar
Model	TSM-PD14
Cell type	Si Multicrystalline
Maximum Power (STC conditions)	320 W
Efficiency (η)	17.5%
Dimensions (h × w × d)	$1960 \times 992 \times 40$ mm^3
Temperature Coefficient of maximum power (α)	−0.41%/K

2. For the wind facility, a vertical-axis wind turbine generator was selected with a nameplate power capacity of 6 kW (similar to the PV facility capacity). Figure 15 shows the VAWT power curve for standard density (ρ_{std} = 1.225 kg/cm^2).

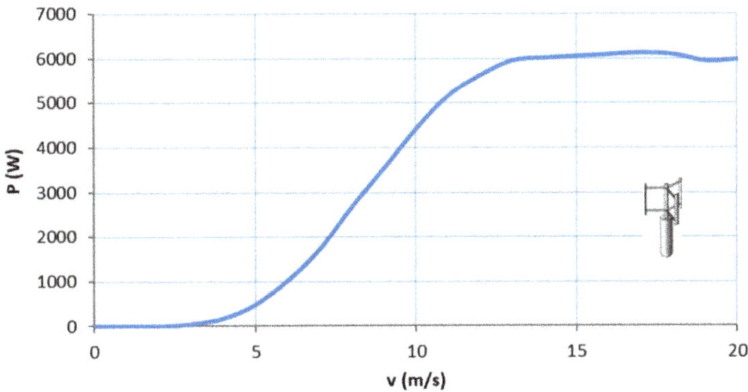

Figure 15. Power curve for the standard air density ρ_{std} = 1.225 kg/cm^2. Source [86].

Figure 16 illustrates the comparison of the matching factor for PV+W hybrid plants ε^{PV+W} versus PV ε^{PV} obtained for facilities with a power capacity of 3.6 kW and 6 kW. In a global context, the adaptation obtained for the 6 kW facility is 19% higher for the method 1, 15% higher for the method 2 and 13% for the method 3. When it is compared with the 3.6 kW, the improvement of the 6 kW facility is higher in all the climate areas.

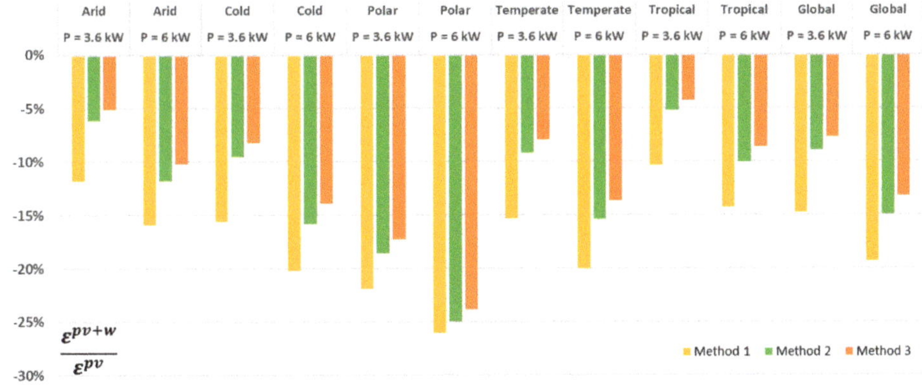

Figure 16. Improvement of the matching factor: ε^{pv+w} (hybrid) over ε^{pv} (PV) facilities for different installed power capacity.

The matching factor for the PV+W hybrid ε^{PV+W} versus wind facilities ε^{W} obtained for the facilities of 3.6 kW and 6 kW is compared in Figure 17. As was obtained for the 3.6 kW facility, the improvement of the matching factor obtained for the 6 kW facility is better when it is compared with the wind facility than when it is compared with the PV facility. The improvement now reaches 32% for method 1, 28% for method 2 and 27% for method 3. The values maintain quite similar ranges in all the climate areas.

The variation of ε^{PV+W} with the multiplication factors introduced in the Chapter 3.1.1. for the 6 kW PV+W hybrid facility is illustrated in the Figure 18. As it can be shown, for the new capacity the methodology also presents a robust performance because the results hardly vary with changes of the irradiation for the three normalisation methods. Once again, the effect of variations in the wind resource is quite limited.

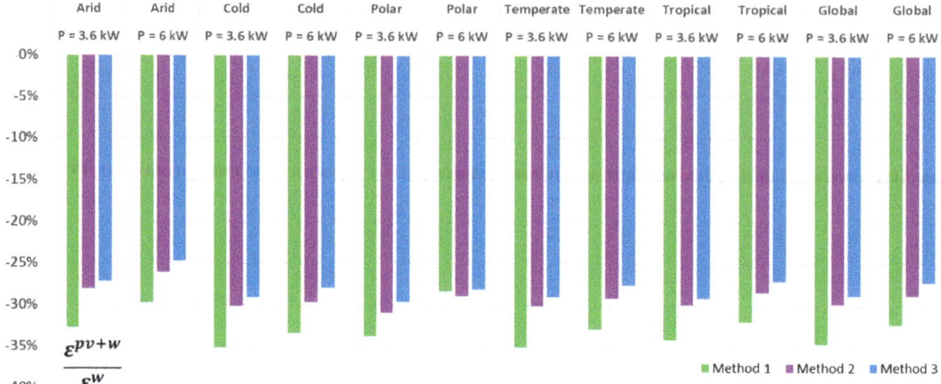

Figure 17. Improvement of the matching factor: ε^{pv+w} (hybrid) over ε^{w} (wind) facilities for different installed power capacity.

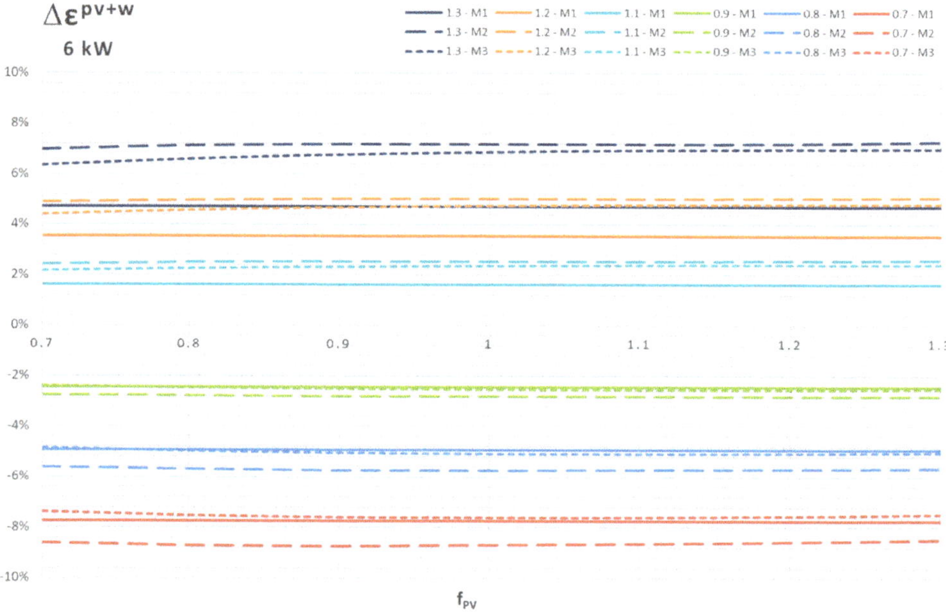

Figure 18. Variation of ε^{PV+W} with the multiplication factors f_{pv} and f_{w}. in a 6 kW power capacity facility (M1, M2 and M3 represent the results obtained by means of application of the normalisation methods 1, 2 and 3, respectively).

4. Discussion

In this paper, we have analysed the behaviour of PV+W hybrid facilities placed in urban areas from the point of view of the adaptability of their generation patterns to the aggregate demand profiles. With this aim, we have designed a novel methodology that includes the definition and calculation of the matching factor (ε) to evaluate and quantify the adaptation level. The novelty of our work is based on three main grounds: (i) the evaluation of supply–demand balance adaptation of PV+W hybrid plants, (ii) the integration of the hybrid plants into an urban environment and (iii) the applicability of the results on a global scale.

The analysis of the generation patterns shows that, in a PV+W hybrid plant where the PV and wind facilities have similar installed power capacity, the PV is always the main contributor in the total energy production in all climate conditions, presenting a global value of 84%, varying from 71% in polar areas to 91% in tropical zones. The main reason for this performance is that the facilities are not placed following a criterion of high-wind-resource location which is common in urban areas.

The results show that PV facilities match demand profiles better than wind energy. The global matching factor obtained for PV ε^{PV} is 0.46 while for wind ε^{W} is 0.6, which means 30% worse adaptation level. The difference, once again, is homogeneous in all climate conditions.

Likewise, hybrid plants adapt better to the demand than when the facilities are independently evaluated. The hybrid plants present ε^{PV+W} in the 0.4 to 0.43 range, depending on the normalisation method used, which means an improvement between 7.7% and 15% in comparison with the adaptation of PV facilities and between 29% and 35% in comparison with wind plants. Once again, the results are homogeneous for all the climate zones.

The proposed methodology has been found robust because the results obtained do not vary substantially with respect to the variation of the solar irradiation or the mean wind speed at the location under study. The methodology also gives comparable results for facilities with different power capacity.

5. Conclusions

An important technical challenge for a massive RES integration is the lack of manageability of the generation to match the demand. A high RES penetration requires the application of measures focused on planning, operation and flexibility of the whole system to respond to the uncertainty and variability in the supply–demand balance in short timescales. These measures present tangible costs to the system.

The results of this study lead us to state that the implementation of PV+W hybrid plants in urban areas would widen the RES integration limits and reduce the cost of high RES penetration because of the improvement of the manageability derived of a better adaptation to the demand profiles.

Additionally, our work gives valuable and quantifiable support to decision-makers to favour RES penetration into the urban environment, which constitutes a perfect example of distributed generation, with the advantages that this type of generation presents for the electric system.

As a result, a massive installation of PV+W hybrid plants would bring benefits for the whole electric system. Therefore, the author's workgroup of the Department of Electrical, Electronic and Control Engineering [87] propose and recommend the implementation of PV+W hybrid plants.

A massive integration into the urban environment presents financial, technical and regulatory barriers. The adequation of existing buildings could require the evaluation of subsidies to avoid financial constraints that could slow down the integration. Moreover, the facilities integrated into the urban environment will be likely owned by consumers (i.e., particulars or small/medium size companies) that could use part of the generation for self-consumption. This situation will require specific legislation to regulate the energy trading and implement technical requirements to avoid negative impact on the distribution networks.

Author Contributions: Conceptualisation: S.C.-R.; methodology: S.C.-R.; software: S.C.-R.; validation: S.C.-R.; formal analysis: S.C.-R.; investigation, S.C.-R.; resources: S.C.-R.; writing—original draft preparation: S.C.-R.; writing—review and editing: A.L.-R., S.C.-R., R.G.-O. and A.C.-S.; supervision: A.C.-S.

Funding: This research received no external funding.

Acknowledgments: The authors would like to thank Laj Cheenikkaparambil Abdullah for his contribution in making this article more readable.

Conflicts of Interest: The authors declare no conflict of interest.

Nomenclature

General

APV	PV area
E	Energy generated (Wh)
e	Normalised energy generated
G	Yearly solar irradiation (insolation) in Wh/m^2 incident on an optimally tilted solar panel
i	Facility type: PV (Photovoltaic), W (Wind) or PV+W (Hybrid)
L	Load electricity demand profile (Wh)
l	Normalised load electricity demand profile
P	Power (W)
PR	Performance ratio of the PV facility
PV	Photovoltaic electricity source
RES	Renewable Energy Source

Subscripts

x	Location
y	Season: S (spring), U (summer), A (autumn), W (winter).
z	Day type: D (weekday), E (weekend).

Greeks

α	Temperature coefficient of maximum power
ε	Matching factor
η	Solar panel efficiency
ρ	Air density

References

1. United Nations. UN Climate Change Conference Paris 2015. Available online: http://www.un.org/sustainabledevelopment/cop21/ (accessed on 9 July 2016).
2. United Nations Framework Convention on Climate Change. Available online: http://unfccc.int/2860.php (accessed on 5 July 2016).
3. United Nations Treaty Collection. Paris Agreement. Available online: https://treaties.un.org/pages/ViewDetails.aspx?src=TREATY&mtdsg_no=XXVII-7-d&chapter=27&lang=en (accessed on 9 July 2016).
4. The US Department of Energy; Office of Energy Efficiency & Renewable Energy. The U.S. Department of Energy (DOE) SunShot Initiative. Available online: http://energy.gov/eere/sunshot/sunshot-initiative (accessed on 25 June 2016).
5. The US Department of Energy; Office of Energy Efficiency & Renewable Energy. The U.S. Department of Energy (DOE) Wind Program. Available online: http://energy.gov/eere/wind/about-doe-wind-program (accessed on 25 June 2016).
6. European Climate Foundation. *Roadmap 2050: A Practical Guide to a Prosperous, Low Carbon Europe*; European Climate Foundation: The Hague, The Netherlands, 2010.
7. International Renewable Energy Agency. RESOURCE. 2018. Available online: http://resourceirena.irena.org/gateway/dashboard/ (accessed on 22 April 2018).
8. REN21. *Renewables 2018 Global Status Report*; REN21 Secretariat: Paris, France, 2018; p. 325.
9. Philipps, S.P.; Kost, C.; Schlegl, S. *Up-to-Date Levelised Cost of Electricity of Photovoltaics*; Fraunhofer Institute for Solar Energy Systems ISE: Freiburg, Germany, 2014.
10. Energy Technology Systems Analysis Programme; International Renewable Energy Agency. *Wind Power. Technology Brief*; International Renewable Energy Agency: Abu Dhabi, UAE, 2016; p. 28.
11. Lazard. In *Levelized Cost of Energy v11*; Lazard: New York, NY, USA, 2017; p. 22.
12. Feldman, D.; Margolis, R.; Boff, D. *Q4 2016/Q1 2017 Solar Industry Update*; U.S. Department of Energy: Oak Ridge, TN, USA, 2017; p. 83.
13. Huld, T.; Jäger Waldau, A.; Ossenbrink, H.; Szabo, S.; Dunlop, E.; Taylor, N. *Cost Maps for Unsubsidised Photovoltaic Electricity*; JRC 91937 Joint Research Centre of the European Commission: Brussels, Belgium, 2014; p. 22.

14. Moné, C.; Hand, M.; Bolinger, M.; Rand, J.; Heimiller, D.; Ho, J. *2015 Cost of Wind Energy Review*; NREL/TP-6A20-66861; National Renewable Energy Laboratory: Golden, CO, USA, 2017; p. 115.

15. International Renewable Energy Agency. *Renewable Power Generation Costs in 2017*; International Renewable Energy Agency: Abu Dhabi, UAE, 2018; p. 160.

16. Cayir Ervural, B.; Evren, R.; Delen, D. A multi-objective decision-making approach for sustainable energy investment planning. *Renew. Energy* **2018**, *126*, 387–402. [CrossRef]

17. Patlitzianas, K.D.; Doukas, H.; Kagiannas, A.G.; Psarras, J. Sustainable energy policy indicators: Review and recommendations. *Renew. Energy* **2008**, *33*, 966–973. [CrossRef]

18. Liu, G.; Li, M.; Zhou, B.; Chen, Y.; Liao, S. General indicator for techno-economic assessment of renewable energy resources. *Energy Convers. Manag.* **2018**, *156*, 416–426. [CrossRef]

19. International Energy Agency. *IEA Energy Technology Perspectives 2017*; International Energy Agency: Paris, France, 2017.

20. International Energy Agency. *Technology Roadmap Wind Energy. 2013 Edition*; International Energy Agency: Paris, France, 2013; p. 63.

21. International Energy Agency. *Technology Roadmap Solar Photovoltaic Energy. 2014 Edition*; International Energy Agency: Paris, France, 2014; p. 20.

22. International Renewable Energy Agency. *REmap: Roadmap for a Renewable Energy Future*; International Renewable Energy Agency: Abu Dhabi, UAE, 2016; p. 172.

23. International Renewable Energy Agency. *Renewable Energy in Cities*; International Renewable Energy Agency: Abu Dhabi, UAE, 2016; p. 64.

24. Karteris, M.; Slini, T.; Papadopoulos, A.M. Urban solar energy potential in Greece: A statistical calculation model of suitable built roof areas for photovoltaics. *Energy Build.* **2013**, *62*, 459–468. [CrossRef]

25. Defaix, P.R.; van Sark, W.G.J.H.M.; Worrell, E.; de Visser, E. Technical potential for photovoltaics on buildings in the EU-27. *Sol. Energy* **2012**, *86*, 2644–2653. [CrossRef]

26. Singh, R.; Banerjee, R. Estimation of rooftop solar photovoltaic potential of a city. *Sol. Energy* **2015**, *115*, 589–602. [CrossRef]

27. Wiginton, L.K.; Nguyen, H.T.; Pearce, J.M. Quantifying rooftop solar photovoltaic potential for regional renewable energy policy. *Comput. Environ. Urban Syst.* **2010**, *34*, 345–357. [CrossRef]

28. Colmenar-Santos, A.; Campíñez-Romero, S.; Pérez-Molina, C.; Mur-Pérez, F. An assessment of photovoltaic potential in shopping centres. *Sol. Energy* **2016**, *135*, 662–673. [CrossRef]

29. Mohajeri, N.; Upadhyay, G.; Gudmundsson, A.; Assouline, D.; Kämpf, J.; Scartezzini, J.-L. Effects of urban compactness on solar energy potential. *Renew. Energy* **2016**, *93*, 469–482. [CrossRef]

30. Song, X.; Huang, Y.; Zhao, C.; Liu, Y.; Lu, Y.; Chang, Y.; Yang, J. An Approach for Estimating Solar Photovoltaic Potential Based on Rooftop Retrieval from Remote Sensing Images. *Energies* **2018**, *11*, 3172. [CrossRef]

31. Polo, M.-E.; Pozo, M.; Quirós, E. Circular Statistics Applied to the Study of the Solar Radiation Potential of Rooftops in a Medium-Sized City. *Energies* **2018**, *11*, 2813. [CrossRef]

32. Moraitis, P.; Kausika, B.; Nortier, N.; van Sark, W. Urban Environment and Solar PV Performance: The Case of the Netherlands. *Energies* **2018**, *11*, 1333. [CrossRef]

33. Toja-Silva, F.; Colmenar-Santos, A.; Castro-Gil, M. Urban wind energy exploitation systems: Behaviour under multidirectional flow conditions—Opportunities and challenges. *Renew. Sustain. Energy Rev.* **2013**, *24*, 364–378. [CrossRef]

34. Araújo, A.M.; de Alencar Valença, D.A.; Asibor, A.I.; Rosas, P.A.C. An approach to simulate wind fields around an urban environment for wind energy application. *Environ. Fluid Mech.* **2013**, *13*, 33–50. [CrossRef]

35. Grant, A.; Johnstone, C.; Kelly, N. Urban wind energy conversion: The potential of ducted turbines. *Renew. Energy* **2008**, *33*, 1157–1163. [CrossRef]

36. Grieser, B.; Sunak, Y.; Madlener, R. Economics of small wind turbines in urban settings: An empirical investigation for Germany. *Renew. Energy* **2015**, *78*, 334–350. [CrossRef]

37. Pagnini, L.C.; Burlando, M.; Repetto, M.P. Experimental power curve of small-size wind turbines in turbulent urban environment. *Appl. Energy* **2015**, *154*, 112–121. [CrossRef]

38. Alexandru-Mihai, C.; Alexandru, B.; Ionut-Cosmin, O.; Florin, F. New Urban Vertical Axis Wind Turbine Design. *INCAS Bull.* **2015**, *7*, 67–76. [CrossRef]

39. Carcangiu, S.; Montisci, A. A building-integrated eolic system for the exploitation of wind energy in urban areas. In Proceedings of the 2012 IEEE International Energy Conference and Exhibition (ENERGYCON), Florence, Italy, 9–12 September 2012; pp. 172–177.

40. Pagnini, L.; Piccardo, G.; Repetto, M.P. Full scale behavior of a small size vertical axis wind turbine. *Renew. Energy* **2018**, *127*, 41–55. [CrossRef]

41. Micallef, D.; van Bussel, G. A Review of Urban Wind Energy Research: Aerodynamics and Other Challenges. *Energies* **2018**, *11*, 2204. [CrossRef]

42. Klima, K.; Apt, J. Geographic smoothing of solar PV: Results from Gujarat. *Environ. Res. Lett.* **2015**, *10*, 104001. [CrossRef]

43. Monforti, F.; Huld, T.; Bódis, K.; Vitali, L.; D'Isidoro, M.; Lacal Arantegui, R. Assessing complementarity of wind and solar resources for energy production in Italy. *A Monte Carlo approach.* **2014**, *63*, 576–586. [CrossRef]

44. Santos-Alamillos, F.J.; Pozo-Vázquez, D.; Ruiz-Arias, J.A.; Lara-Fanego, V.; Tovar-Pescador, J. Analysis of Spatiotemporal Balancing between Wind and Solar Energy Resources in the Southern Iberian Peninsula. *J. Appl. Meteorol. Climatol.* **2012**, *51*, 2005–2024. [CrossRef]

45. Widen, J. Correlations Between Large-Scale Solar and Wind Power in a Future Scenario for Sweden. *IEEE Trans. Sustain. Energy* **2011**, *2*, 177–184. [CrossRef]

46. Zhang, H.; Cao, Y.; Zhang, Y.; Terzija, V. Quantitative synergy assessment of regional wind-solar energy resources based on MERRA reanalysis data. *Appl. Energy* **2018**, *216*, 172–182. [CrossRef]

47. Prasad, A.A.; Taylor, R.A.; Kay, M. Assessment of solar and wind resource synergy in Australia. *Appl. Energy* **2017**, *190*, 354–367. [CrossRef]

48. Huber, M.; Dimkova, D.; Hamacher, T. Integration of wind and solar power in Europe: Assessment of flexibility requirements. *Energy* **2014**, *69*, 236–246. [CrossRef]

49. Solomon, A.A.; Kammen, D.M.; Callaway, D. Investigating the impact of wind–solar complementarities on energy storage requirement and the corresponding supply reliability criteria. *Appl. Energy* **2016**, *168*, 130–145. [CrossRef]

50. Jurasz, J.; Beluco, A.; Canales, F.A. The impact of complementarity on power supply reliability of small scale hybrid energy systems. *Energy* **2018**, *161*, 737–743. [CrossRef]

51. The US Department of Energy, Office of Electricity Delivery & Energy Reliability. *The Potential Benefits of Distributed Generation and the Rate-Related Issues That May Impede Its Expansion*; U.S. Department of Energy: Oak Ridge, TN, USA, 2015.

52. Abdmouleh, Z.; Gastli, A.; Ben-Brahim, L.; Haouari, M.; Al-Emadi, N.A. Review of optimization techniques applied for the integration of distributed generation from renewable energy sources. *Renew. Energy* **2017**, *113*, 266–280. [CrossRef]

53. Ruiz-Romero, S.; Colmenar-Santos, A.; Gil-Ortego, R.; Molina-Bonilla, A. Distributed generation: The definitive boost for renewable energy in Spain. *Renew. Energy* **2013**, *53*, 354–364. [CrossRef]

54. Mai, T.; Wiser, R.; Sandor, D.; Brinkman, G.; Heath, G.; Denholm, P.; Hostick, D.J.; Darghouth, N.; Schlosser, A.; Strzepek, K. *Exploration of High-Penetration Renewable Electricity Futures*; NREL/TP-6A20-52409-1; National Renewable Energy Laboratory: Golden, CO, USA, 2012.

55. International Energy Agency. *System Integration of Renewables. An update on Best Practice*; International Energy Agency: Paris, France, 2018.

56. Batalla-Bejerano, J.; Trujillo-Baute, E. Impacts of intermittent renewable generation on electricity system costs. *Energy Policy* **2016**, *94*, 411–420. [CrossRef]

57. Delarue, E.; Van Hertem, D.; Bruninx, K.; Ergun, H.; May, K.; Van den Bergh, K. *Determining the Impact of Renewable Energy on Balancing Costs, Back Up Costs, Grid Costs and Subsidies*; Report for Adviesraad Gas en Elektriciteit—CREG; CREG: Brussels, Belgium, 2016.

58. Hirth, L.; Ueckerdt, F.; Edenhofer, O. Integration costs revisited—An economic framework for wind and solar variability. *Renew. Energy* **2015**, *74*, 925–939. [CrossRef]

59. Shang, Y. Resilient Multiscale Coordination Control against Adversarial Nodes. *Energies* **2018**, *11*, 1844. [CrossRef]

60. Spanish National Statistics Centre. *Buildings and Population Census*; Spanish National Statistics Centre: Madrid, Spain, 2019.

61. Spanish Ministry of Development. *Buildings Construction Licences*; Spanish Ministry of Development: Madrid, Spain, 2019.

62. Eurostat. CensusHub2. 2019. Available online: https://ec.europa.eu/CensusHub2/query.do?step=selectHyperCube&qhc=false (accessed on 20 March 2019).

63. U.S. Census Bureau. American FactFinder. Annual Estimates of Housing Units for the United States, Regions, Divisions, States, and Counties: April 1, 2010 to July 1, 2017. 2017 Population Estimates. Available online: https://factfinder.census.gov/faces/tableservices/jsf/pages/productview.xhtml?pid=PEP_2018_PEPANNRES&src=pt (accessed on 20 March 2019).

64. U.S. Energy Information Administration. *Commercial Buildings Energy Consumption Survey (CBECS)*; U.S. Energy Information Administration: Washington, DC, USA, 2016.

65. Red Eléctrica de España, S.A. Esios. System Operator Information System. Available online: https://www.esios.ree.es/en/analysis/10004?vis=1&start_date=31-03-2016T00%3A00&end_date=31-03-2016T23%3A50&compare_start_date=30-03-2016T00%3A00&groupby=minutes10&compare_indicators=545,544 (accessed on 2 April 2018).

66. ENTSO-E. *Electricity in Europe 2017. Synthetic Overview of Electric System Consumption, Generation and Exchanges in 34 European Countries*; ENTSO-E: Brussels, Belgium, 2018; p. 20.

67. ENTSO-E. *Monthly Hourly Load Values*; ENTSO-E: Brussels, Belgium, 2019.

68. U.S. Energy Information Administration. *U.S. Electric System Operating Data*; U.S. Energy Information Administration: Washington, DC, USA, 2019.

69. *Meteotest Genossenchaff Meteonorm*, 7th ed. Available online: https://meteonorm.com/t (accessed on 10 January 2018).

70. Köppen, W.; Geiger, R. Das geographische System der Klimate. In *Handbuch der Klimatologie*; Köppen, W., Geiger, R., Eds.; Verlag von Gebrüder Borntraeger: Berlin, Germany, 1936; Volume I, p. 44.

71. Kottek, M.; Grieser, J.; Beck, C.; Rudolf, B.; Rubel, F. World Map of the Köppen-Geiger climate classification updated. *Meteorol. Z.* **2006**, *15*, 259–263. [CrossRef]

72. University of Veterinary Medicine Vienna. Institute for Veterinary Public Health. World Maps of Köppen-Geiger Climate Classification. Available online: http://koeppen-geiger.vu-wien.ac.at/ (accessed on 20 June 2016).

73. Duffie, J.A.; Beckman, W.A.; Worek, W.M. *Solar Engineering of Thermal Processes*, 4th ed.; John Wiley & Sons, Inc.: Hoboken, NJ, USA, 2013; p. 928.

74. International Technology Roadmap for Photovoltaic. *ITRPV 2017 Results*; VDMA: Francfurt, Germany, 2018; p. 71.

75. Trina Solar. Products. PC14. 72-Cell Utility Module. Available online: http://www.trinasolar.com/uk/product/PC14.html (accessed on 15 May 2016).

76. Reich, N.H.; Mueller, B.; Armbruster, A.; van Sark, W.G.J.H.M.; Kiefer, K.; Reise, C. Performance ratio revisited: Is PR > 90% realistic? *Prog. Photovolt. Res. Appl.* **2012**, *20*, 717–726. [CrossRef]

77. Dierauf, T.; Growitz, A.; Kurtz, S.; Becerra Cruz, J.L.; Riley, E.; Hansen, C. *Weather-Corrected Performance Ratio*; NREL/TP-5200-57991; National Renewable Energy Laboratory: Golden, CO, USA, 2013.

78. V-AIR. Vertical Axis Wind Turbine VisionAir5. Available online: http://www.visionairwind.com/wp-content/uploads/2018/08/VisionAIR5.pdf (accessed on 6 April 2019).

79. Carta González, J.A.; Calero Pérez, R.; Colmenar-Santos, A.; Castro Gil, M.-A. *Centrales de Energías Renovables: Generación Eléctrica con Energías Renovables*; Pearson Educación, S.A.: Madrid, Spain, 2009; p. 728.

80. PJM. Data Viewer. Load. Available online: https://dataviewer.pjm.com/dataviewer/pages/public/load.jsf (accessed on 2 April 2018).

81. Northwest Power Pool. Geographic NWPP Aggregated Wind Generation and Load Data. Available online: http://www.nwpp.org/our-resources/NWPP-Reserve-Sharing-Group/Geographic-NWPP-Aggregated-Wind-Generation-and-Load-Data (accessed on 2 April 2018).

82. Midcontinent Independent System Operator. I. Market Reports. Available online: https://www.misoenergy.org/markets-and-operations/market-reports/#nt=%2FMarketReportType%3ASummary%2FMarketReportName%3ADaily%20Regional%20Forecast%20and%20Actual%20Load%20(xls)&t=10&p=0&s=MarketReportPublished&sd=desc (accessed on 2 April 2018).

83. Coordinador Eléctrico Nacional. Operación Real. Available online: https://sic.coordinador.cl/informes-y-documentos/fichas/operacion-real/ (accessed on 1 April 2018).

84. Australian Energy Market Operation. Electricity Load Profiles—Aggregated Price and Demand Data—Historical. Available online: https://www.aemo.com.au/Electricity/National-Electricity-Market-NEM/Data-dashboard#aggregated-data (accessed on 1 April 2018).
85. Trina Solar. Products. PD14. 72-Cell Utility Module. Available online: http://static.trinasolar.com/sites/default/files/ES_TSM_PD14_datasheet_B_2017_web.pdf (accessed on 21 April 2018).
86. Kingspan Renewable Technologies. KW6 Wind Turbine. Available online: https://www.kingspan.com/gb/en-gb/products/renewable-technologies/wind-energy/kw6-wind-turbine (accessed on 21 April 2018).
87. National Distance Education University. Department of Electrical, Electronic and Control Engineering. Available online: http://www2.uned.es/personal/antoniocolmenar/publicaciones.htm (accessed on 1 May 2016).

Article

An Energy Potential Estimation Methodology and Novel Prototype Design for Building-Integrated Wind Turbines

Oscar Garcia [1], Alain Ulazia [2,*], Mario del Rio [1], Sheila Carreno-Madinabeitia [3] and Andoni Gonzalez-Arceo [1]

[1]   ROSEO start-up, University of the Basque Country (UPV/EHU), Alda. Urkijo, 48013 Bilbao, Spain; ogarcia076@ikasle.ehu.eus (O.G.); mdelrio009@ikasle.ehu.eus (M.d.R.); agonzalez440@ikasle.ehu.eus (A.G.-A.)

[2]   Department of NE and Fluid Mechanics, University of the Basque Country (UPV/EHU), Otaola 29, 20600 Eibar, Spain

[3]   Meteorology Area, Energy and Environment Division, TECNALIA R&I, 01001 Vitoria-Gasteiz, Spain; sheila.carreno@tecnalia.com

*   Correspondence: alain.ulazia@ehu.eus

Received: 13 April 2019; Accepted: 23 May 2019; Published: 27 May 2019

Abstract: ROSEO-BIWT is a new Building-Integrated Wind Turbine (BIWT) intended for installation on the edge of buildings. It consists of a Savonius wind turbine and guiding vanes to accelerate the usual horizontal wind, together with the vertical upward air stream on the wall. This edge effect improves the performance of the wind turbine, and its architectural integration is also beneficial. The hypothetical performance and design configuration were studied for a university building in Eibar city using wind data from the ERA5 reanalysis (European Centre for Medium-Range Weather Forecasts' reanalysis), an anemometer to calibrate the data, and the actual small-scale behavior in a wind tunnel. The data acquired by the anemometer show high correlations with the ERA5 data in the direction parallel to the valley, and the calibration is therefore valid. According to the results, a wind speed augmentation factor of three due to the edge effect and concentration vanes would lead to a increase in working hours at the rated power, resulting annually in more than 2000 h.

Keywords: building integrated wind turbine; savonius; ERA5; anemometer; calibration

---

1. Introduction

In general terms, the market for small wind turbines is currently growing, although the sector of small wind turbines intended for installation in buildings is increasing at a lower rate. According to the World Wind Energy Association (WWEA) [1], the installation of small wind turbines will increase by around 12% annually in the 2015–2020 period. The good economic profitability of small wind turbines and the consistency of technological advancement are determinant factors that explain the growth of the small wind turbine market. On the other hand, in the last several years, research is increasingly being focused on the development of different technologies that help minimize the energy consumption of buildings. This philosophy is known as nZEB (nearly Zero-Energy Building) [2], and it is included in the EU 2010/31/CE directive related to the energy efficiency of buildings. After 2018, every new public building should be constructed in accord with this regulation and, after 2020, every new building should be compliant.

The goal is to maximize energy efficiency and reduce the use of primary energy derived from fossil resources so that the required energy demand can be met by renewable sources. In this sense, mini wind technology, which involves generating energy with wind turbines of 100 kW or less to cover an area smaller than 200 m^2, can play a very important role. However, some technological challenges, such as the vibrations, the generated noise levels, and the device's aesthetic and architectonic integration, are yet to be fully solved.

Nevertheless, these devices have many advantages:

1. They can work as standalone devices, so they can provide energy in isolated locations without a connection to the electric grid.
2. They work in distributed micro-generation mode, thus minimizing energy losses due to transport and distribution. These devices generate energy at a site that is close to the final user, thus dramatically reducing the need for electric infrastructures.
3. Furthermore, it can be combined with photovoltaic energy in hybrid installations to enable the optimal use and management of shared electric accumulators.

The recent developments in wind energy for urban environments have inspired different types of Building-Integrated Wind Turbine (BIWT) projects. For example, in London, Strata SE1 is a tall building with 43 floors that will include three wind turbines with diameters of 9 m on the roof of the structure. These wind turbines will be used to meet the building's lighting demand [3].

On a smaller scale, there are a lot of projects that include Horizontal-Axis Wind Turbines (HAWTs) integrated with buildings, as well as Vertical-Axis Wind Turbines (VAWTs). These projects are focused on integrating wind turbines with existing buildings. Thus, these buildings were not previously designed to accelerate air streams, unlike the World Trade Center of Baharein [4] or the mentioned Strata **SE1** building. According to this post-integration trend, building-integrated wind turbines are being implemented in strategic locations to capture the acceleration of air streams that are produced because of different geometries. In this sense, the most interesting locations are the upper and lateral edges of a building, especially the former because it is at a reasonable distance from homes.

Nowadays, there are several ongoing projects working to develop an optimal system that harnesses wind energy in urban environments. Most of them have concluded that wind turbines located in obstacle-free environments are not adequate for urban environments because of the urban turbulent flow, which can present a relevant turbulence intensity on the superior edges of the buildings [5,6]. For that reason, HAWT devices, which usually exhibit good performances with laminar flows, perform poorly in urban environments, in addition to their generation of noise as high as 200 dB within a radius of 500 m [7]. Conversely, VAWTs play an essential role in generating wind energy in urban areas since their performance is not much affected by turbulent flows, and they tend to be noiseless [8]. Additionally, the VAWT has a lower cut-in speed than HAWT and a larger or even unlimited cut-off speed, ensuring longer operating times [9–11]. Although the power coefficient is lower, the design is simpler and the manufacturing process is easier to carry out.

Along these lines, the existing urban wind energy potential has encouraged researchers to develop a proper methodology for wind energy estimation in urban environments [12]. The use of anemometers at specific locations can be combined with advanced computational simulations of buildings situated in complex urban terrains using CFD (Computational Fluid Dynamics). In this way, wind energy potential estimation using reanalysis and meteorological mesoscale models, which is a well-known offshore and onshore method and also developed by the authors [13,14], can be complemented with different back-end tools.

In this work, the authors present the design of a Savonius drag-driven turbine that is intended for integration into buildings. The proposed turbine is called ROSEO-BIWT, which has been specially designed to work in urban environments. The wind in urban areas is characterized by its turbulence, thus it is important to take advantage of low-speed air streams. The germinal project of ROSEO won the first award in the EDP-RENEWABLE UNIVERSITY CHALLENGE 2017 [15], and the members of the project have now created a university start-up called ROSEO. Although it is typically used as a vertical-axis turbine, ROSEO-BIWT is formed by a Savonius turbine in a horizontal position and concentration vanes that accelerate the air streams by the Venturi effect (see Section 2.2). These types of vanes are usually called PAGVs (Power Augmentation Guiding Vanes) [16–18]. The proposed turbine was also designed to be easily architectonically integrated. This was the case for the design proposed by Park et al. [19], in which several Savonius turbines were incorporated into the facade of a building at different heights to take advantage of the vertical currents created by the wind on the walls of the building.

The Savonius wind turbine is a drag-based device, unlike the majority of turbines, which are lift-based. This particular aspect allows for low noise levels and few vibrations, and these factors are very important in building installations [20,21]. The PAGV increases the wind speed as the catching area grows, resulting in a system that is able to start at wind speeds of about 1 m/s, thus ensuring a great number of energy-producing hours. Furthermore, energy generation continues no matter how high the wind speed is.

This paper proceeds as follows: a possible location for the installation, which was established using ERA5 data, is presented. ERA5 is a powerful tool for global atmospheric analysis that is updated in real time (see Section 2.1). The authors also installed an anemometer on the roof of their university to calibrate the wind data for a period of eight months against ERA5 (Section 2.1.2). In this way, an empirical method for the estimation of wind energy potential on buildings with a low computational cost will be developed in subsequent work, as discussed in Section 3.5. Finally, a preliminary small-scale experiment was developed for a wind tunnel with a small Savonius and different configurations of the PAGV (Sections 2.2 and 2.3). The authors finish this work with some relevant conclusions and a future outlook of possible research directions. The qualitative methodology used here can be considered within the scope of analogical reasoning and model construction [22].

2. Data and Methodology

2.1. Data and Location

2.1.1. Anemometers and ERA5

The university building of Eibar (Engineering School of Gipuzkoa) was selected (longitude: 2.946° W; latitude: 43.258° N) to demonstrate a method for formulating a preliminary estimation of energy production using quantile-matching calibration versus a cup anemometer installed on the roof. Figure 1 shows a satellite view of the engineering school and the position of two anemometers installed on two buildings. After eight months of data acquisition, Anemometer 1 showed the best correlations with the ERA5 reanalysis, and its dataset was used for the calibration and energy estimation procedure.

The ERA5 reanalysis, ECMWF's most recent atmospheric reanalysis, covers the second half of the 20th century and this century [23]. For this study and the calibration, 40 years of data were used (from 1979 to 2018), because the complete reanalysis is not yet available. The correlation between ERA5 and the anemometers was computed within their period of intersection (from June 2018 to February 2019). These data include atmospheric and oceanic variables, and are an appropriate tool for estimating wind energy potential [24] offshore and onshore. In this study, ERA5 hourly data with a resolution of $0.3° \times 0.3°$ was used.

Figure 1. Selected buildings and anemometers 1 and 2 on the roof.

2.1.2. Quantile-Mapping Calibration

The cup anemometers installed on the buildings of the University of Basque Country in Eibar enabled the development of a preliminary calibration methodology based on quantile-matching techniques that were used previously by the authors for wind energy and wave energy [25–27]. In the scientific literature, different calibration or bias correction techniques have been developed and compared for the analysis of several parameters, such as temperature and precipitation (see [28–30]). Data from models and reanalysis are compared with observations. In the present study, a simple but effective statistical procedure based on quantile mapping was used.

For this approach, several other terms can be found in the literature: "probability mapping" [31], "quantile-quantile mapping" [32,33], "statistical downscaling" [34], and "histogram equalization" [35]. With this general approach, empirical quantile-mapping bias correction was applied to calibrate ERA5 versus an anemometer in the building. In [36], the same procedure was used for estimating wind energy trends. To summarize, this method of calibration or bias correction is fundamentally statistical, and the idea is to match values with the same quantile in two empirical probability distributions: the one to be calibrated (ERA5), and the one that is the basis for the calibration (anemometer). Figure 2 illustrates the main aspects of this calibration procedure, including the intersection periods and the concept of applying the transference function.

For this paper, the authors obtained an eight-month, 10-min data series, which was filtered every 6 h to match the ERA5 reanalysis for a 10-year period (1-h time resolution, in this case). Thus, there were around 34,500 cases in the anemometer time series and around 87,600 cases in the ERA5 series. Taking 1-h data for both series in the intersection period resulted in 5390 cases, from which the correlation was measured and the subsequent calibration transference function was generated.

Having determined the average wind speed $\overline{U}$ after calibration on the corresponding facade and considering the typical shape parameter of the Weibull distribution (Rayleigh distribution, $k = 2$), the corresponding scale parameter can be obtained:

$$c = \overline{U}/\Gamma(1 + 1/k). \tag{1}$$

Then, the cumulative distribution function and the fraction of time between two wind speeds are determined:

$$F(U) = 1 - exp(-(U/c)^k) \tag{2}$$

and the augmentation factor AF of the PAGV (the ratio between the outlet and inlet free wind speed) can be incorporated into the c parameter [9] because it is proportional to the average wind speed $\overline{U}$.

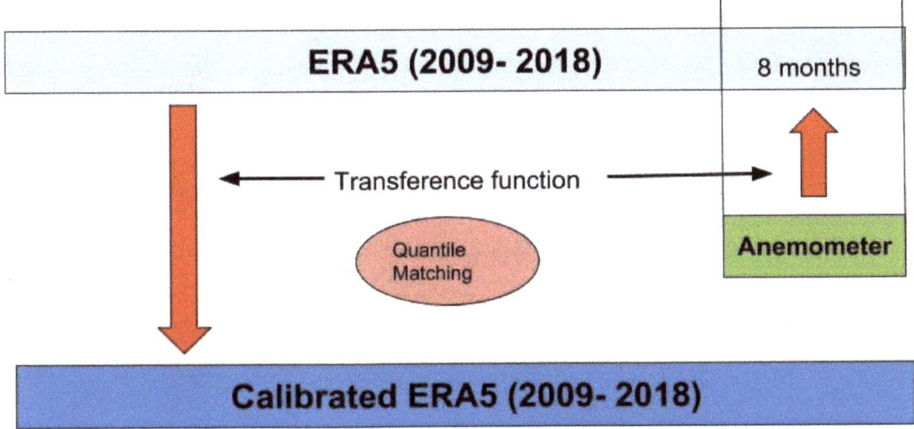

Figure 2. Calibration procedure and periods of the ERA5 and anemometer data.

The value of AF can only be based on a virtual definition of the outlet velocity of the flux, since the complex interaction between the diffusive flux in the exterior part of the vanes and the motion of the rotor do not permit a simplistic application of the Venturi effect according to the relation between the capture width of the free wind U and the outlet width. However, the optimum tip speed ratio (TSR_{opt}) at which the power coefficient C_p is maximized is directly related to the outlet effective velocity (U_{out}), because it is well known that $TSR_{opt0} \approx 1/3$ for a drag turbine without augmentation techniques [9]. Due to the Magnus and lift effects in the Savonius rotor, this value can reach 0.4–0.5. Therefore, the augment of TSR_{opt} should be similar to AF considering an effective U_{out} at the position of the rotor in relation to the blade tip speed V_{tip}. Being V_{tip}^A the tip speed in the augmented rotor and TSR^A the tip speed ratio in the augmented rotor, the hypothesis is that the TSR should be the same for the conventional Savonius and for the augmentation technique if the outlet velocity $AF \cdot U$ is the reference wind speed:

$$TSR = \frac{V_{tip}}{U} = \frac{V_{tip}^A}{AF \cdot U} \Rightarrow V_{tip}^A = AF \cdot V_{tip} \tag{3}$$

However, the TSR with augmentation TSR^A should be defined with respect to the free wind speed: $\frac{V_{tip}^A}{U}$. Therefore,

$$TSR^A = AF \cdot TSR \Rightarrow AF = \frac{TSR^A}{TSR} \tag{4}$$

Consequently, AF can be also computed using the ratio of the TSR with augmentation versus the TSR without augmentation.

On the other hand, experiment using nozzles by Shika et al. [37] have shown that AF can be 4 or even 5 measuring directly U_{out} at the position of the rotor for U between 0.6 and 0.9 m/s. This relevant increment for low free wind speed is very interesting for our purpose, since reducing significantly the cut-in speed of the rotor. Furthermore, these AFs ensure a great quantity of working hours at rated power, as shown below.

2.2. ROSEO-BIWT Design

2.2.1. The Location on the Upper Edge of the Building

The effect of wind against buildings has been largely studied by the architectural sector for the purpose of studying the dynamic loads generated by air streams. Because of their work, there is considerable knowledge about the behavior of wind in urban environments. Much of the information has been obtained through experiments with scale models and CFD simulations, similar to the depiction generated by the authors in Figure 3, which was re-created based on the CFD simulation in [38].

Figure 3. Re-creation on the basis of Mertens [38] for wind acceleration over the windward upper edge of a building.

Most of the studies that have analyzed the behavior of air streams around buildings agree that the upper edge of the windward face of a building has great wind energy potential. This is because the wind has to surround an object. The effect is even more intense when the building is taller and when the wind direction is perpendicular to the building facade. For example, in a five-story building, the wind velocity increases by 1.2 times at the windward edge [38].

According to the CFD simulations of Balduzzi et al. [5,6], the wind speed increment at the edge can be between 10% and 30%, but the turbulence intensity increases considerable. This is not the worst inconvenience for Savonius turbines, since it is demonstrated that, when turbulence increases, the separation of boundary layer takes place on the lower side of returning blade of the rotor reducing the negative torque [39,40].

Areas of high turbulent intensity create more frequent and stronger gusts [9], but the inertia of a relatively long Savonius rotor (high aspect ratio between the length of the axis and the diameter) can keep the rotation of the turbine without relevant variations. Additionally, it is demonstrated that, as a consequence of blade tips, aerodynamic losses are reduced in Savonius turbines with high aspect ratios [41].

2.2.2. Savonius Turbine

Because of the above-discussed wind behavior, our ROSEO-BIWT's Savonius axis is positioned horizontally along the superior edge of the building. There have been several recent studies on the performance of the Savonius turbine. Mohamed et al. [42] improved the performance using plates to eliminate the negative torque in the returning blade. They carried out tests for a two-bladed and a three-bladed wind turbine, and, in both cases, they improved the power coefficient (C_p) of the wind turbine by up to 27%, with 15% being the typical value.

Apart from these intrinsic improvements, some engineers have developed the mentioned PAGV systems to accelerate air streams. Shikha et al. [43] increased the wind speed by 3.7 times in an experiment using a specific well-studied nozzle. Additionally, Altan et al. [44] studied the influence of the inclination angle of the plates as well as their length. In these experiments, they found that, when the longitude of the PAGV increased, the power also increased. Thus, the important consideration in their study was the relationship between the diameter of the rotor and the length of the PAGV. They even obtained a C_p of 38.5%. Other types of PAGVs, referred to as omnidirectional, reached a C_p of 48%, implying an increase of 240% relative to a Savonius rotor without a PAGV system.

In terms of longitude and diameter, the Savonius rotor studied in [45] performed best with an aspect ratio of 6:1. Similarly, Park et al. [19] tested different kinds of Savonius rotors, and they discovered that the best design was a six-bladed rotor. Therefore, for our purpose, a similar rotor with these proportions was chosen for the initial test period.

2.2.3. The Final Design

Park et al. [19] developed the idea of using a larger facade surface to generate energy by installing a lot of Savonius rotors at different heights while also using PAGVs to improve the performance of the wind turbines. The system that they proposed is similar to a ventilated facade. It is important to emphasize that they wanted to capture the vertical air streams that are generated on the windward side of the building, as in our case. However, they used parallel vanes in the facade with a small concentration angle; in our case, the upper edge is used to augment the concentration angle and capture not only the vertical stream on the facade but also the horizontal component of the wind. Furthermore, the background of the Savonius rotor is free on the edge of the building: this is an important aspect that is not encountered in turbines installed in the facade. Although the influence of this aspect is out of the scope of this study, it is an obvious aerodynamic advantage.

Another innovation is that the proposed turbine can be installed in existing buildings: it is not a design intended only for new buildings. To summarize, ROSEO-BIWT shows good architectural integration in existing buildings and high economic viability due to the simplicity of the design.

Figure 4 shows the ROSEO-BIWT design. This is a schematic perspective that does not take into account the influence of the angle between the two PAGVs; the angle can be adapted for other positions of the vanes.

Figure 4. ROSEO-BIWT design.

The PAGV areal ratio between the entrance and exit of the air is 4:1, and a similar AF is expected to result from a first simplistic calculus due to the Venturi effect. In any case, as mentioned, the complex interaction between the outlet wind speed and the rotor motion deviates this a priori value of $AF = 4$.

In their seminal work about a curtain design to increase the performance of a Savonius turbine, Altan et al. [46] determined that the best angles for the capture of wind in their curtain design are $15°$ for the superior vane and $45°$ for the inferior one that obstructs the negative torque. The authors corroborated the same influence of the inferior vane in the laboratory (see Section 3.1). Figure 5 shows the dimensions for this optimum design with curtains. It is considered a unit of capture width at the inlet, and a geometrical relation of 4:1 for the inlet width (0.25, therefore diameter of 0.50) versus the outlet width. The aspect ratio is six considering the results of Roy and Saha [45]: $3 = 0.5 \times 6$. These proportions can be established between a capture width of one and two meter; within this size, ROSEO device is manipulable for a worker on the roof in the implementation process and for O&M issues, without the need of a crane .

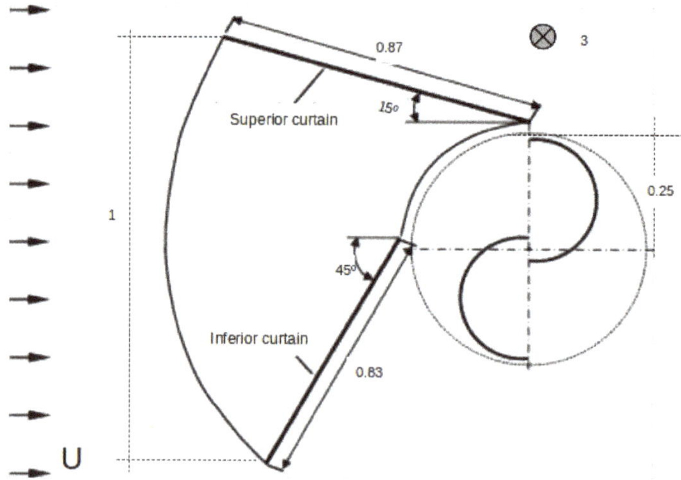

Figure 5. Adapted figure of the optimum design using curtain vanes [46].

2.3. Experiments in the Wind Tunnel

Although there are results provided by previous studies, in the following sections, the authors describe the general experimental methodology that is being developed. The experimental model construction is proposed by referencing previous findings and design procedures [22]. These are the main steps:

1. First, according to the literature, the augmentation factor of the wind speed on the edge of the buildings is around 1.2. Wind speed augmentation is the result of the union between the usual horizontal component and the vertical component.
2. Then, the previous augmentation factor should be multiplied by the new increment AF provided by the vanes. These factors will be measured for different wind speeds in the wind tunnel of the university using a small-scale model of a building with curtain-type vanes (see Figure 5) and a rotor or 2 cm diameter.
3. A similar experiment will be performed for a real Savonius with one inferior vane and will be critically compared with other studies.
4. Finally, the Weibull distribution at the location obtained by the previously described calibration methodology will be applied to the measured power curve that includes AF. Thus, the amount of hours at rated power due to this augmentation will be an interesting parameter about energy production.

Table 1 describes the main characteristics of the above-mentioned wind tunnel. Figure 6 shows the wind tunnel and the installation of a PAGV and a real Savonius rotor. It should be mentioned that the disposition of the vane below the limit of the rotor's horizontal axis obstructs the negative torque and, simultaneously, accelerates the stream in the opened drag side above the axis. As mentioned, this aerodynamic effect has been properly documented in previous reviews about the performance of the Savonius rotor [39,41,47].

Table 1. Characteristics of the wind tunnel.

Length; diameter	2 m; 630 mm
Measuring system	Pitot tubes, an ultrasonic anemometer, and air pressure transducers
Range of wind speed	0–13 m/s
Materials	Structure of aluminum and dome of polycarbonate
Control panel	Potentiometer for the regulation of wind speed, rpm, and torque
Generator	*maxon RE motor* 65 mm, Graphite Brushes, 250 Watt [48]
Data acquisition	Variable resistor with measurement of voltage, intensity, and power

The augmentation may be even higher because of the corner effect of our design. However, until now, these preliminary measurements have only been performed with low values of steady wind speed without considering some important effects, such as the blockage ratio of the tunnel [49,50]. However, the the influence of AF is important at these low wind speeds below the rated power, because it can ensure a sufficient wind speed above the rated wind speed at the outlet of the concentration vanes.

On the other hand, Figure 7 shows the small-building, the guiding vanes with 3:1 inlet/outlet relation, and the six-bladed rotor of 2 cm diameter. Here, the objective is to create an anemometer that is able to capture the wind on the entire outlet area of the vanes. According to our previous experiments, measurements with Pitot tubes result in great fluctuations due to small displacements or inclination deviation in such a narrow area. In this case, the electrical motor is a *maxon DCX06M EB KL 6V* of 0.529 W. The speed constant is of 3060 $\text{min}^{-1}\,\text{V}^{-1}$ having a direct way to compute the angular velocity in function of the voltage. Additionally, there is also a constant speed–torque relation of 36,600 $\text{min}^{-1}\,\text{mNm}^{-1}$. This characteristic is important because it allows computing the increment of the torque due to the increment of the speed.

Figure 6. The Savonius turbine inside the wind tunnel with the inferior vane.

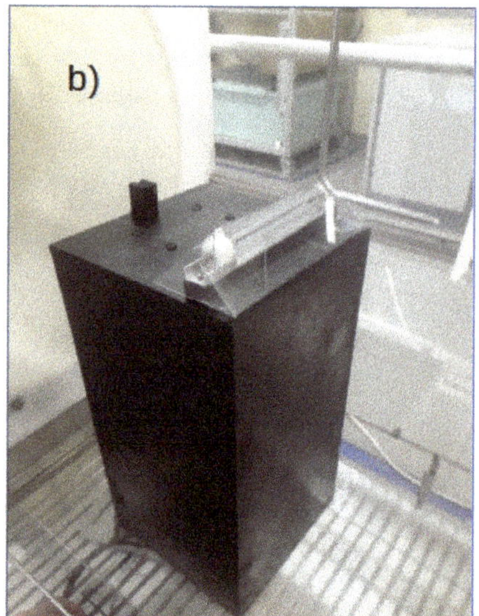

Figure 7. The small building model (**a**) without the vanes; and (**b**) with the vanes.

3. Results

3.1. Effect of the PAGV in the Real Savonius

AF of around 2 has been corroborated in the experiments of the wind tunnel. This obviously depends on the angle and the length of the vanes and the exact position of the Pitot tube, but this value of AF can be obtained with a suitable disposition of the vanes. However, this measurement is strongly influenced by the exact position and size of the Pitot tube, and future works should study the behavior of AF for higher and more turbulent wind speeds.

Although a maximum outlet/inlet width ratio of 1.4 can be obtained due to the lack of space in the tunnel with the Savonius rotor in the center, the instantaneous power measured for different wind speeds and different angles of the vane in Figure 6 gives a coherent result for this augmentation and subsequent power. Figure 8 shows this behavior with the vane at 30°, 45° and 70° with respect to the horizontal. The pilot test was developed without the vane and the negative-torque wall (NT wall) was applied with the vane in vertical position, obstructing the negative torque's drag. It should be noted that, removing the negative torque, the power is doubled and the other cases (30°, 45° and 70°) also remove the negative drag.

The results in the curves of Figure 8 show the best working condition for the vane at 45°, in which the captured power almost triples the pilot test power at each wind speed. Being the geometrical augment relation of 1.4, and $1.4^3 \approx 3$, the power also keeps the typical proportionality relation with U^3 for AF, and, again, a constant AF equal to the geometrical relation is deduced. This fact establishes an important particular case for an hypothetical law that should be demonstrated: for the adequate vane angles, the estimation of power production can be performed using $AF \times U$ as the input wind speed for any free wind speed U.

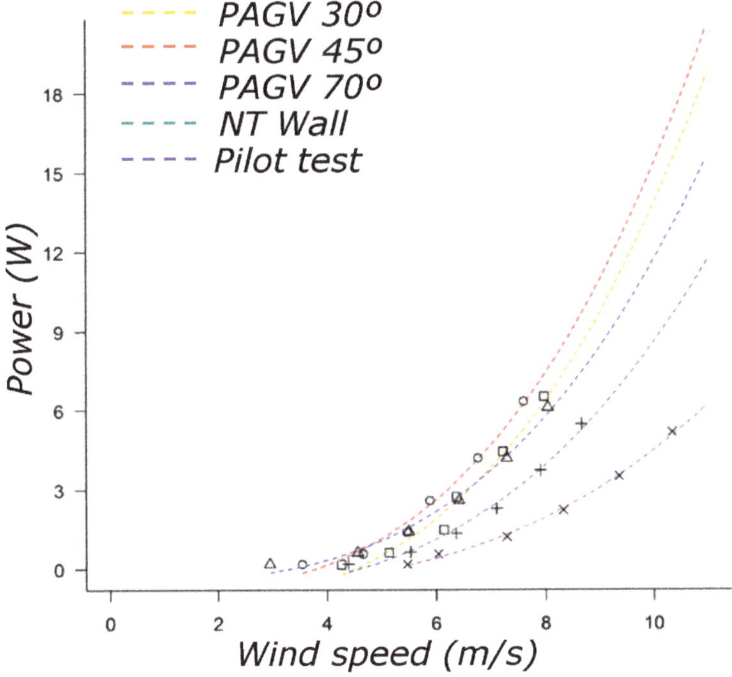

Figure 8. Power production versus wind speed for different positions of the vane and the pilot test without the vane.

The behavior of C_p vs. TSR was also studied and is presented in Table 2 for the optima with which AF can be estimated. The presence of the vane increments TSR_{opt} from 0.5 to around 1 with $AF \approx 2$. The best case in power augmentation (almost three times) for $AF = 2.2$ is measured for the vane at 45°. These results are totally coherent with previous works using different augmentation techniques that obtain $TSR_{opt} \approx 1$ compared to a value of 0.5 for a conventional Savonius. In these cases, $C_{p,max}$ is also two or even three times higher thanks to the vanes, deflectors, curtains or other kinds of concentration configurations [44,51]. Additionally, as mentioned above, Altan et al. also already showed for their curtain type augmentation technique that the inferior vane should be at 45° to optimize the energy capture [52].

Table 2. Optimum C_p, and corresponding TSR, AF and power increment for the Savonius turbine in the wind tunnel for the pilot text, negative torque vertical wall, and different angles of the vane.

Experiment	$C_{p,max}(\%)$	TSR_{opt}	AF
PAGV30	17.1	1.01	2.0
PAGV45	19.2	1.10	2.2
PAGV70	16.1	0.94	1.9
NT Wall	11.6	0.68	1.4
Pilot test	6.5	0.50	-

3.2. Augmentation Factor in the Small-Scale Building Model

In this case, TSR for each wind speed is measured instead of the TSR_{opt}. Consequently, the ratio of the $TSRs$ with (TSR_v) and without (TSR_0) the vane should be corrected according to the rotor speed, since the torque is incremented with the speed. Figure 9 shows these results: both TSRs, their ratio, and the corrected ratio that equals AF. This correction factor is established by the increment relation between U and the cut-in wind speed for the pilot experiment (5 m/s). A logical step in the procedure considering the constant speed–torque relation of the DC generator, since both TSRs (therefore, the rotor speeds) are practically linear with respect to U and also fulfill the same increment relation. Thus, AF is between 2.5 and 3, a very relevant result given the fact that the inlet–outlet width relation is 3:1. The 4:1 relation of the initial configuration could not be installed yet, due to the sensible construction details of the small model. Because of this delicate structure, the fabrication process of which has been very laborious, the free wind speed range in the tunnel has been kept below 10 m/s.

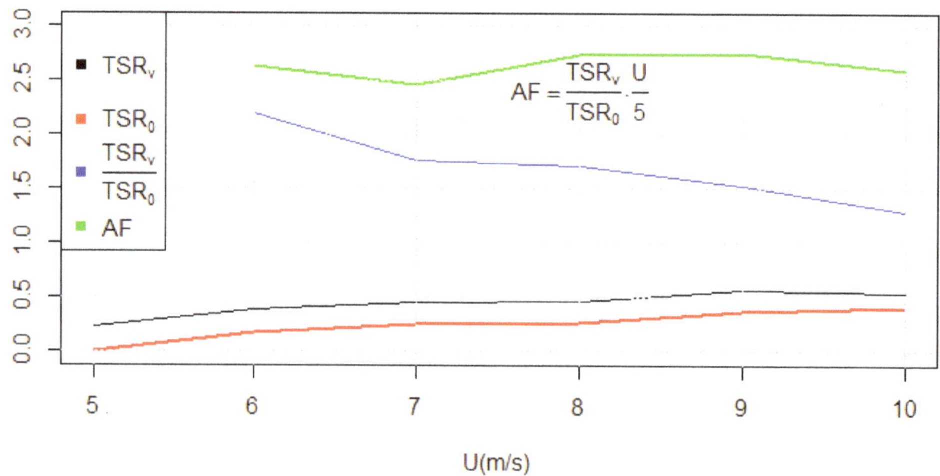

Figure 9. TSRs, the ratio, and the corrected ratio for the rotor speed with and without the vane.

3.3. Wind Rose around the Building

The ERA5 grid around Eibar city is shown in Figure 10 with the ERA5 points in blue. The building on which the anemometer is located is marked in red. The nearest grid point, at a distance of 2.48 km, was chosen to perform the calibration.

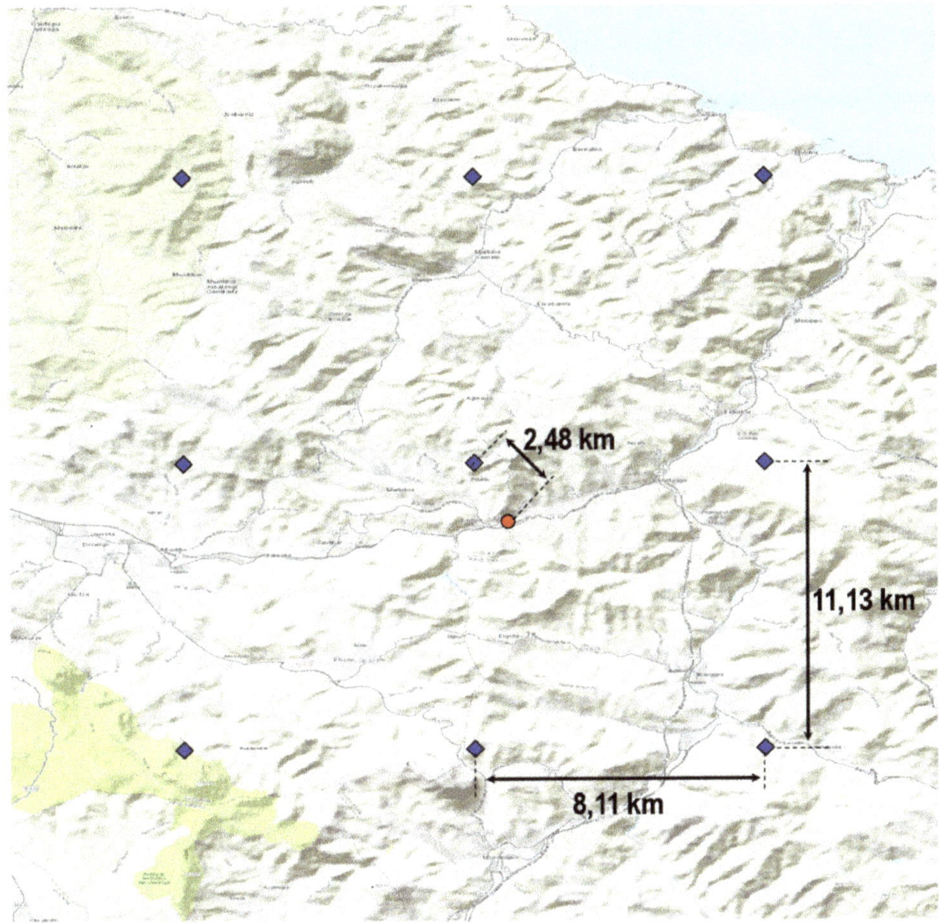

Figure 10. Nearest ERA5 grid points (blue) around the study point (red).

With the wind rose representing the nearest ERA5 grid point and the anemometer (see Figure 11), it is easy to realize that the ERA5 data have to be calibrated to make an appropriate estimation. ERA5's wind rose shows a strong predominant direction toward the northwest, as it is well-known that the climate of the Basque Country is highly related to the behavior of geostrophic winds [14]. This predominant direction is perpendicular to the valley in Eibar, and it is clearly diminished by the roughness of the terrain and the obstacle of the mountains in the anemometer's data. In fact, Eibar is an industrial city with a population of 20,000 in a deep valley surrounded by mountains that are around 600 m from where the River Deba opens toward the northeast direction.

Thus, this big difference could be explained by the shape and direction of the valley in which Eibar is located. It shows the need for field measurements and indicates that a good calibration methodology must use atmospheric reanalysis to study wind potential in places such as cities and deep valleys, which have high surface roughness. Furthermore, the valley direction determines not only the calibration direction for energy estimation purposes, but it also defines which of the facades of the building should be selected for the implementation of the BIWT.

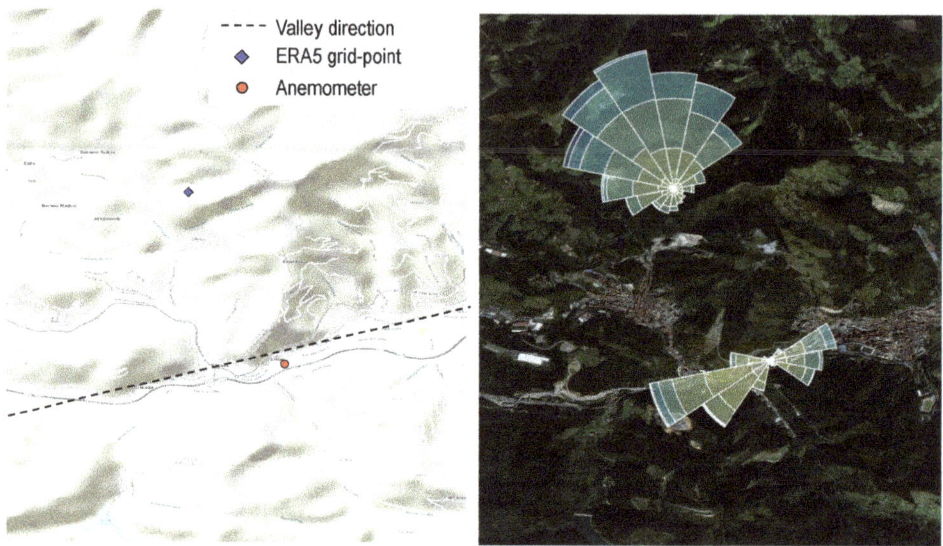

Figure 11. Details of the location of the anemometer and the nearest ERA5 grid point (**left**). Representation of the corresponding wind roses of ERA5 and anemometer data (**right**).

3.4. Comparison between ERA5 and the Anemometer

These data should be established at a referential height using the log law and the roughness of urban environments [9]. The ERA5 grid-point height is 411 m, thus both datasets should be established at the same height, which is the anemometer's height in this case, since it is the observation.

According to usual considerations in the wind energy sector, the roughness (z_0) of the urban terrain is around 1–10 m. Roughness is used to apply the logarithmic law of vertical wind shear,

$$\frac{U(z)}{U(z_r)} = \frac{Ln(z/z_0)}{Ln(z_r/z_0)},$$ (5)

which results in a correction factor between 0.86 and 0.77 for a wind speed at a height of 178 m in ERA5. In terms of the speed of reference at its original height, the authors calculated the following:

$$U(178) = \frac{Ln(178/1)}{Ln(411/1)}U(411) = 0.86 \times U(411); \frac{Ln(178/10)}{Ln(411/10)}U(411) = 0.77 \times U(411)$$ (6)

A correction factor of 0.86 was used prior to the calibration method, which is based on quantile mapping. However, first, the correlation between ERA5 and the anemometer had to be directionally studied, mainly in the direction of the valley line. Furthermore, anemometer data had been previously filtered using advanced filters in meteorology, such as temporal checks, persistence tests, and climate-based range tests [53], which were implemented in the R programming language by the authors [54].

Figure 12 shows a time series of a week in June 2018 when the Pearson's correlation between ERA5 and the anemometer was very high (around 0.95); the wind direction vectors are illustrated above each time point that shows a strong westerly component. The parallel patterns shown by the wind speed series are obvious in the graph. These examples verify the quality of the anemometer's data since their results are comparable to the reliable ERA5 data in the predominant direction line established by the valley (southwest–northeast). In the first days of this week, a cut-off low occurred in the Bay of Biscay, and it caused strong wind and a large amount of precipitation in that area. During the following days, without the influence of the cut-off low, wind moving in the north direction was observed. This is a global-scale synoptic situation that is easier to detect by the ERA5 model than local setups. Therefore, a high correlation between the observed data and ERA5 data was confirmed.

Additionally, the approximation to observation of the corrected ERA5 signal resulting from the application of the log law is clear in the time series using the 0.86 factor, but it is not enough to totally correct the general overestimation presented by ERA5. Although an extreme correction for a high roughness $z_0 = 10$ m with a factor of 0.77 would strongly reduce this overestimation (Equation (6)), the usual roughness values for urban environments are kept in this graph.

This example is an extraordinary case, but, if all the cases of wind between the south and east were selected during the study period, a good correlation of 0.70 would finally be obtained. This validation was therefore enough to justify the calibration in this directional range, from which the corresponding building facade will be selected for capturing wind energy.

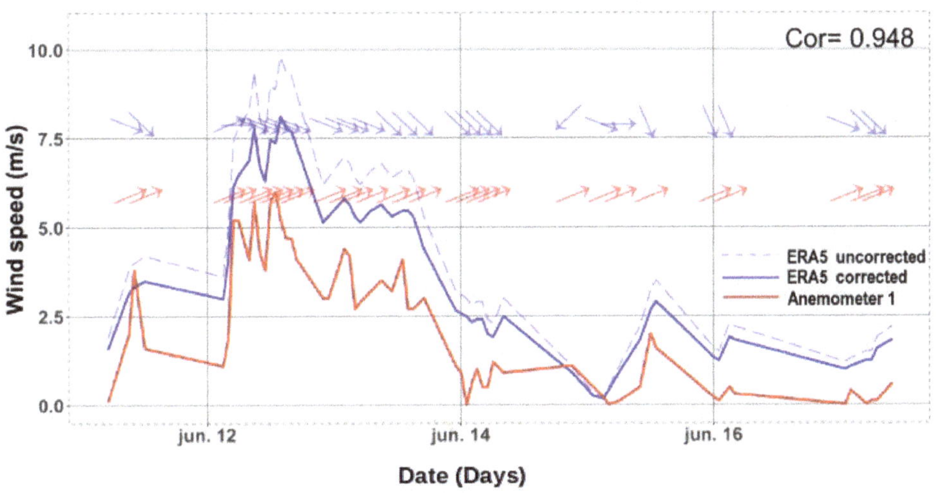

Figure 12. Wind speed of nearest ERA5 grid point (blue) and anemometer (red) during a week of June. In addition, wind direction vectors are represented in the graph.

3.5. Estimation of the Energy Potential

On the basis of the resource assessment results of the wind potential in buildings and the described wind tunnel experiments, a general methodology is presented that also references previous results from the scientific literature to estimate the annual energy production (AEP) of ROSEO-BIWT:

1. According to Mertens [38] and the initial experiments with our small-scale building in the wind tunnel, the wind increases its velocity by 20% at the upper edge of a typical building.
2. The simplest PAGVs have increased the wind speed by four times, with a corresponding increase in C_p to a value as high as 0.37 [47]. Although higher values can be obtained with wider entrances, the authors will use an AF of 4 for the estimation, although there is also a 20% augmentation due to the additional architectonic acceleration at the upper edge.

3. Taking into account the wind rose in Figure 11, the authors only considered the wind data of ERA5 for the valley direction and for our turbine on the corresponding facade.

4. $AF \approx 3$ has been corroborated by our small-scale building with PAGVs for different wind speed values in the wind tunnel. Although the optimum vane angle experiment has not yet been developed, the first test results are consistent with values reported in the literature.

5. $AF \approx 2$ has been corroborated by the real Savonius with the inferior vane.

6. The analyses of the wind resource in the open direction of the valley and the corresponding facade yield a wind speed histogram or Weibull distribution that can be applied to the power curve of the turbine with AF.

7. For the first estimation presented here, the working time at rated power due to the increment of wind speed using PAGVs has been computed.

Although the results of the comparison and the calibration of data are very relevant, the objective of this paper is mainly methodological, and a preliminary estimation of AEP should be made using a well-known device. Thus, for the estimation of generated power, a commercial Savonius model (SeaHawk-PACWIND) was used: the rated power is of 1.1 kW, the rated wind speed is 17.9 m/s, the cut-in wind speed is 3.1 m/s, the cut-off without a given limit is a drag device, and the swept area is 0.92 m^2 [55].

When the pure ERA5 wind speed distribution was considered for the best facade, after the quantile-matching calibration using the anemometer data, these are the preliminary results:

- The average wind speed is 4.2 m/s, and the shape factor k is around 2, depending on the angle range in the predominant direction, i.e., southwest (see Equations (2) and (1)).

- The turbine's working hours per year in the interval of rated wind speed (above 17.9 m/s) can be computed if the cumulative density function $F(U)$ (Equation (2)) is applied to c, which results in the following working hours:

$$[1 - F(17.9)] \times 365.25 \times 24 = 160 \tag{7}$$

- Therefore, AEP is $160 \times 1.1 = 170$ kWh at the rated power; it is a very small value since the working hours of a profitable turbine should be around 2000 h per year.

- However, multiplying the scale parameter c by values between 2 and 4 ($AF = 2$ is the value obtained in the laboratory using only one inferior vane and 4 the maximum expected value according to the mentioned literature) and keeping the typical value of $k = 2$, the total augmentation factor AF from the PAGVs and the edge effect increases the AEP and working hours. Figure 13 shows the annual working hours (WH_{hours}) at rated power in function of the average wind speed $\overline{U}$ of the site for different factors: $AF = 2; 2.5; 3; 3.5; 4$.

At low annual average wind speed of 3 m/s, the maximum $AF = 4$ can produce 2000 h at rated power. At $\overline{U} = 4$ m/s, AF between 2.5 and 3 is necessary to ensure the 2000 h. At $\overline{U} = 5$ m/s, the minimum $AF = 2$ obtained with only the inferior vane (Figure 5) is almost sufficient. At high $\overline{U}$s, an $AF = 3.5$ or 4 implies 75% of the time at rated power.

Annual working hours at rated power

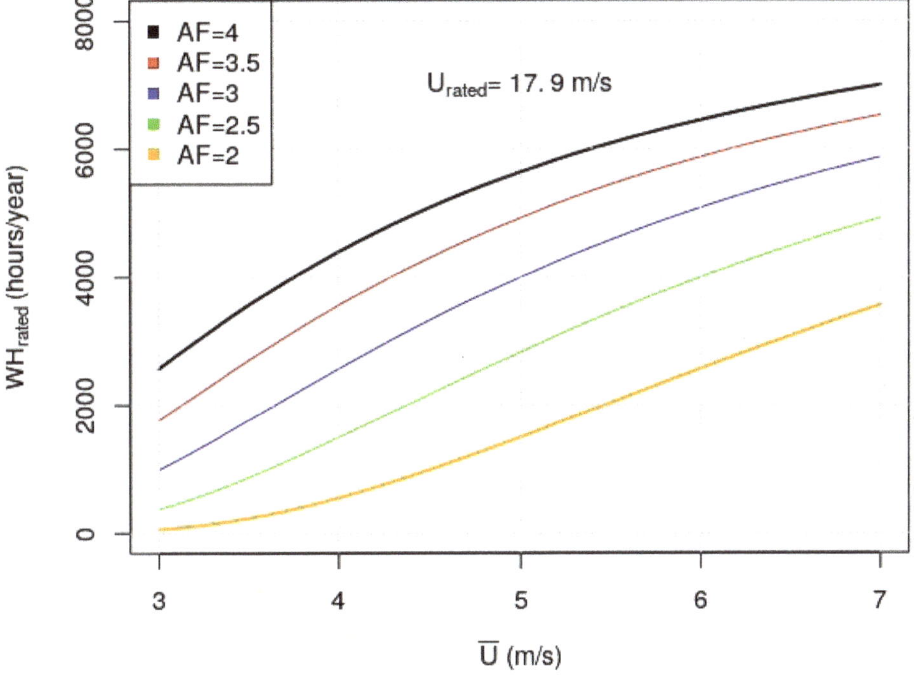

Figure 13. Annual working hours at rated power versus $\overline{U}$ for different AF values.

Given the strong directionality presented by the anemometer's wind rose (Figure 11), two ROSEO-BIWTs installed in opposite facades of the building that are perpendicular to the valley would capture almost all of the mentioned hours because winds that are perpendicular to the valley are infrequent.

4. Conclusions and Future Outlook

An integral methodology with preliminary results is presented for a new type of BIWT. The preliminary results include the energy potential estimation, measurements of small-scale building aerodynamic effects, and the influence of PAGVs. In the future, an AEP increase of 20% via PAGVs at the edge of the buildings must be demonstrated using a real prototype of ROSEO-BIWT at the edge of the building in Eibar. For that, the building in the city of Eibar will be used in the Bizia Lab project of the University of Basque Country, together with the previous wind tunnel experiments for the mentioned small-scale building with PAGVs and short Savonius prototype.

The anemometer was installed on the roof; with the new data provided by ERA5 for the nearest grid point, the identification of the best facade and the corresponding wind distribution were obtained following the methodology described in this paper applied to a longer period. This methodology will be relevant when the one-year period has elapsed. These preliminary results and the methodological discussion developed to date encourage us to implement future refinements of ROSEO-BIWT and the related wind energy estimation methodology.

If the building edge effect and the PAGVs produce a wind speed augmentation of AF, our general mathematical proof for working hours at the rated power shows that the hours without the augmentation can be considerably incremented. This is a very important general rule for turbines with vanes, as shown in Figure 13. Furthermore, the values of AF between two and four are coherent with the literature and the experimental results, even under-valued, since the augment of the free wind in the edge of the building is not considered. The edge effect augment factor of 1.2 documented by the literature and the higher inlet-outlet width relationship other type of PAGVs could increase the overall AF.

Additionally, a novel validation method for anemometers developed by the authors in a recent study for wind farms [56] will be very beneficial since it enables the comparison and combination of data from more than one anemometer installed on the roof of the building. This allows us to consider both the zonal and meridional components in a single comparison score.

Future experiments in the wind tunnel with a small-scale building and the PAGVs will be carried out to obtain the optimum value of the angle between the vanes and the augmentation factor for different wind speeds within the operating range of the turbine. The augmentation factors measured with an interval of 0.5 m/s within this range, together with the measurement of the power curve of the longitudinal profile of the Savonius with the same step, will allow us to apply the corresponding histogram distribution of the corrected and calibrated wind to the augmented power curve. However, it is expected that future energy production results will be similar to the values presented here.

Finally, it should be emphasized that lacking the ability to change the viscosity of the air in the tunnel is an important inconvenience for small-scale building experiments. In the future, a more advanced tunnel with the ability to change the pressure and temperature is necessary, together with a parallel validation of the results using CFD simulations of the edge effect of the building with PAGVs.

Author Contributions: Conceptualization, A.U., O.G., and M.d.R.; Methodology, A.U., O.G., and M.d.R.; Software, A.U., O.G., M.d.R., S.C.-M., and A.G.-A.; Validation, O.G.; Investigation, A.U., O.G., and A.G.-A.; Writing—Original Draft Preparation, A.U. and O.G.; Writing—Review and Editing, all authors; Supervision, all authors; Project Administration, A.U.; and Funding Acquisition, A.U.

Funding: The research leading to these results was carried out in the framework of the Programme Campus Bizia Lab EHU (Campus Living Lab) with a financial grant from the Office of Sustainability of the Vice-Chancellorship for Innovation, Social Outreach and Cultural Activities of the University of the Basque Country (UPV/EHU). This programme is supported by the Basque Government. We acknowledge also the availability given by the School of Engineering of Gipuzkoa-Eibar in the University of Basque Country, the EDP-Renewable awards in which we obtained the main award in September 2017, the Youth Enterprise Grant of UPV/EHU, and the project GIU17/02 of EHU/UPV. All computations and representations of this work were developed using the programming language R [54].

Conflicts of Interest: The authors declare no conflict of interest.

Abbreviations

The following abbreviations are used in this manuscript:

CFD	Computational Fluid Dynamics
BIWT	Building-Integrated Wind turbine
O&M	Operation and maintenance
PAGV	Power Augment Guiding Vane
PDF	Probability Density Function
AEP	Annual Energy Production
AF	Augmentation factor
c	Weibull's scale parameter
C_p	Power Coefficient
$C_{p,max}$	Maximum Power Coefficient
k	Weibull's shape parameter
TSR	Tip Speed Ratio
TSR_{opt}	Optimum TSR where C_p is maximum
TSR_{opt}^A	Optimum TSR with augmentation techniques

$\overline{U}$	Average wind Speed
U_p	Wind speed in the prototype
U_m	Wind speed in the model
U_{rated}	Rated wind speed
V_{tip}	Blade tip speed
WT_{rated}	Annual working hours at rated power
z_0	Roughness of the Terrain
z_r	Reference height

References

1. WWEA. *WWEA Released Latest Global Small Wind Statistics*; WWEA: Bonn, Germany, 2018.
2. Chastas, P.; Theodosiou, T.; Bikas, D.; Kontoleon, K. Embodied energy and nearly zero energy buildings: A review in residential buildings. *Procedia Environ. Sci.* **2017**, *38*, 554–561. [CrossRef]
3. Bogle, I. Integrating wind turbines in tall buildings. *CTBUH J.* **2011**, *4*, 30–33.
4. Smith, R.F.; Killa, S. Bahrain World Trade Center (BWTC): The first large-scale integration of wind turbines in a building. *Struct. Des. Tall Spec. Build.* **2007**, *16*, 429–439. [CrossRef]
5. Balduzzi, F.; Bianchini, A.; Ferrari, L. Microeolic turbines in the built environment: Influence of the installation site on the potential energy yield. *Renew. Energy* **2012**, *45*, 163–174. [CrossRef]
6. Balduzzi, F.; Bianchini, A.; Gentiluomo, D.; Ferrara, G.; Ferrari, L. Rooftop siting of a small wind turbine using a hybrid BEM-CFD model. In *Research and Innovation on Wind Energy on Exploitation in Urban Environment Colloquium*; Springer: Riva del Garda , Italy, 2017; pp. 91–112.
7. Oerlemans, S.; Sijtsma, P.; López, B.M. Location and quantification of noise sources on a wind turbine. *J. Sound Vib.* **2007**, *299*, 869–883. [CrossRef]
8. Toja-Silva, F.; Lopez-Garcia, O.; Peralta, C.; Navarro, J.; Cruz, I. An empirical–heuristic optimization of the building-roof geometry for urban wind energy exploitation on high-rise buildings. *Appl. Energy* **2016**, *164*, 769–794. [CrossRef]
9. Manwell, J.F.; McGowan, J.G.; Rogers, A.L. *Wind Energy Explained: Theory, Design and Application*; John Wiley & Sons: Chichester, UK, 2010.
10. Dilimulati, A.; Stathopoulos, T.; Paraschivoiu, M. Wind turbine designs for urban applications: A case study of shrouded diffuser casing for turbines. *J. Wind Eng. Ind. Aerodyn.* **2018**, *175*, 179–192. [CrossRef]
11. Hyams, M. 20-Wind energy in the built environment. In *Metropolitan Sustainability*; Zeman, F., Ed.; Woodhead Publishing Series in Energy; Woodhead Publishing: Cambridge, UK, 2012; pp. 457–499.
12. Arteaga-López, E.; Ángeles-Camacho, C.; Bañuelos-Ruedas, F. Advanced methodology for feasibility studies on building-mounted wind turbines installation in urban environment: Applying CFD analysis. *Energy* **2019**, *167*, 181–188. [CrossRef]
13. Ulazia, A.; Sáenz, J.; Ibarra-Berastegui, G.; González-Rojí, S.J.; Carreno-Madinabeitia, S. Using 3DVAR data assimilation to measure offshore wind energy potential at different turbine heights in the West Mediterranean. *Appl. Energy* **2017**, *208*, 1232–1245. [CrossRef]
14. Ulazia, A.; Saenz, J.; Ibarra-Berastegui, G. Sensitivity to the use of 3DVAR data assimilation in a mesoscale model for estimating offshore wind energy potential. A case study of the Iberian northern coastline. *Appl. Energy* **2016**, *180*, 617–627. [CrossRef]
15. Edp-Renewable University Challenge Awards. Available online: https://www.edpr.com/en/news/2017/09/28/edpr-celebrates-ninth-edition-university-challenge-awards-spain (accessed on 28 September 2017).
16. Chong, W.; Pan, K.; Poh, S.; Fazlizan, A.; Oon, C.; Badarudin, A.; Nik-Ghazali, N. Performance investigation of a power augmented vertical axis wind turbine for urban high-rise application. *Renew. Energy* **2013**, *51*, 388–397. [CrossRef]
17. Chong, W.; Naghavi, M.; Poh, S.; Mahlia, T.; Pan, K. Techno-economic analysis of a wind–solar hybrid renewable energy system with rainwater collection feature for urban high-rise application. *Appl. Energy* **2011**, *88*, 4067–4077.

[CrossRef]

18. Tong, C.W.; Zainon, M.; Chew, P.S.; Kui, S.C.; Keong, W.S.; Chen, P.K. Innovative Power-Augmentation-Guide-Vane Design of Wind-Solar Hybrid Renewable Energy Harvester for Urban High Rise Application. In Proceedings of the 2010 Physics Education Research Conference, Portland, OR, USA, 21–22 July 2010; Volume 1225, pp. 507–521.
19. Park, J.; Jung, H.J.; Lee, S.W.; Park, J. A new building-integrated wind turbine system utilizing the building. *Energies* **2015**, *8*, 11846–11870. [CrossRef]
20. Zemamou, M.; Aggour, M.; Toumi, A. Review of savonius wind turbine design and performance. *Energy Procedia* **2017**, *141*, 383–388. [CrossRef]
21. Kim, S.; Cheong, C. Development of low-noise drag-type vertical wind turbines. *Renew. Energy* **2015**, *79*, 199–208. [CrossRef]
22. Ulazia, A. Multiple roles for analogies in the genesis of fluid mechanics: How analogies can cooperate with other heuristic strategies. *Found. Sci.* **2016**, *21*, 543–565. [CrossRef]
23. Hersbach, H. *The ERA5 Atmospheric Reanalysis*; AGU Fall Meeting Abstracts: San Francisco, CA, USA, 2016.
24. Olauson, J. ERA5: The new champion of wind power modelling? *Renew. Energy* **2018**, *126*, 322–331. [CrossRef]
25. Ulazia, A.; Penalba, M.; Ibarra-Berastegui, G.; Ringwood, J.; Saénz, J. Wave energy trends over the Bay of Biscay and the consequences for wave energy converters. *Energy* **2017**, *141*, 624–634. [CrossRef]
26. Penalba, M.; Ulazia, A.; Ibarra-Berastegui, G.; Ringwood, J.; Sáenz, J. Wave energy resource variation off the west coast of Ireland and its impact on realistic wave energy converters' power absorption. *Appl. Energy* **2018**, *224*, 205–219. [CrossRef]
27. Ulazia, A.; Penalba, M.; Rabanal, A.; Ibarra-Berastegi, G.; Ringwood, J.; Sáenz, J. Historical Evolution of the Wave Resource and Energy Production off the Chilean Coast over the 20th Century. *Energies* **2018**, *11*, 2289. [CrossRef]
28. Teutschbein, C.; Seibert, J. Bias correction of regional climate model simulations for hydrological climate-change impact studies: Review and evaluation of different methods. *J. Hydrol.* **2012**, *456*, 12–29. [CrossRef]
29. Watanabe, S.; Kanae, S.; Seto, S.; Yeh, P.J.F.; Hirabayashi, Y.; Oki, T. Intercomparison of bias-correction methods for monthly temperature and precipitation simulated by multiple climate models. *J. Geophys. Res. Atmos.* **2012**, *117*. [CrossRef]
30. Lafon, T.; Dadson, S.; Buys, G.; Prudhomme, C. Bias correction of daily precipitation simulated by a regional climate model: A comparison of methods. *Int. J. Climatol.* **2013**, *33*, 1367–1381. [CrossRef]
31. Block, P.; Souza Filho, F.; Sun, L.; Kwon, H.H. A streamflow forecasting framework using multiple climate and hydrological models. *J. Am. Water Resour. Assoc.* **2009**, *45*, 828–843. [CrossRef]
32. Boé, J.; Terray, L.; Habets, F.; Martin, E. Statistical and dynamical downscaling of the Seine basin climate for hydro-meteorological studies. *Int. J. Climatol.* **2007**, *27*, 1643–1655. [CrossRef]
33. Sun, F.; Roderick, M.L.; Lim, W.H.; Farquhar, G.D. Hydroclimatic projections for the Murray-Darling Basin based on an ensemble derived from Intergovernmental Panel on Climate Change AR4 climate models. *Water Resour. Res.* **2011**, *47*, W00G02. [CrossRef]
34. Piani, C.; Haerter, J.O.; Coppola, E. Statistical bias correction for daily precipitation in regional climate models over Europe. *Theor. Appl. Climatol.* **2010**, *99*, 187–192. [CrossRef]
35. Rojas, R.; Feyen, L.; Dosio, A.; Bavera, D. Improving pan-European hydrological simulation of extreme events through statistical bias correction of RCM-driven climate simulations. *Hydrol. Earth Syst. Sci.* **2011**, *15*, 2599–2620. [CrossRef]
36. Bett, P.E.; Thornton, H.E.; Clark, R.T. Using the Twentieth Century Reanalysis to assess climate variability for the European wind industry. *Theor. Appl. Climatol.* **2015**, *127*, 1–20. [CrossRef]
37. Shikha.; Bhatti, T.; Kothari, D. Wind energy conversion systems as a distributed source of generation. *J. Energy Eng.* **2003**, *129*, 69–80. [CrossRef]
38. Mertens, S. Wind Energy in the Built Environment: Concentrator Effects of Buildings. Doctoral Thesis, Delf University (TUDelf), Delft, The Netherlands, 2006; ISBN 0906522-35-8
39. Akwa, J.V.; Vielmo, H.A.; Petry, A.P. A review on the performance of Savonius wind turbines. *Renew. Sustain. Energy Rev.* **2012**, *16*, 3054–3064. [CrossRef]

40. Alom, N.; Saha, U.K. Four decades of research into the augmentation techniques of Savonius wind turbine rotor. *J. Energy Resour. Technol.* **2018**, *140*, 050801. [CrossRef]

41. El-Askary, W.; Nasef, M.; Abdel-Hamid, A.; Gad, H. Harvesting wind energy for improving performance of Savonius rotor. *J. Wind Eng. Ind. Aerodyn.* **2015**, *139*, 8–15. [CrossRef]

42. Mohamed, M.; Janiga, G.; Pap, E.; Thévenin, D. Optimization of Savonius turbines using an obstacle shielding the returning blade. *Renew. Energy* **2010**, *35*, 2618–2626. [CrossRef]

43. Bhatti, T.; Kothari, D. A new vertical axis wind rotor using convergent nozzles. In Proceedings of the Large Engineering Systems Conference on Power Engineering, Montreal, QC, Canada, 7–9 May 2003; pp. 177–181.

44. Altan, B.D.; Atılgan, M. A study on increasing the performance of Savonius wind rotors. *J. Mech. Sci. Technol.* **2012**, *26*, 1493–1499. [CrossRef]

45. Roy, S.; Saha, U.K. Review on the numerical investigations into the design and development of Savonius wind rotors. *Renew. Sustain. Energy Rev.* **2013**, *24*, 73–83. [CrossRef]

46. Altan, B.D.; Atılgan, M. The use of a curtain design to increase the performance level of a Savonius wind rotors. *Renew. Energy* **2010**, *35*, 821–829. [CrossRef]

47. Wong, K.H.; Chong, W.T.; Sukiman, N.L.; Poh, S.C.; Shiah, Y.C.; Wang, C.T. Performance enhancements on vertical axis wind turbines using flow augmentation systems: A review. *Renew. Sustain. Energy Rev.* **2017**, *73*, 904–921. [CrossRef]

48. Maxon Motor. Available online: https://www.maxonmotor.com (accessed on 15 March 2019).

49. Usman, M.; Hanif, A.; Kim, I.H.; Jung, H.J. Experimental validation of a novel piezoelectric energy harvesting system employing wake galloping phenomenon for a broad wind spectrum. *Energy* **2018**, *153*, 882–889. [CrossRef]

50. Choi, C.K.; Kwon, D.K. Wind tunnel blockage effects on aerodynamic behavior of bluff body. *Wind Struct. Int. J.* **1998**, *1*, 351–364. [CrossRef]

51. Emmanuel, B.; Jun, W. Numerical study of a six-bladed Savonius wind turbine. *J. Sol. Energy Eng.* **2011**, *133*, 044503. [CrossRef]

52. Altan, B.D.; Atılgan, M. An experimental and numerical study on the improvement of the performance of Savonius wind rotor. *Energy Conv. Manag.* **2008**, *49*, 3425–3432. [CrossRef]

53. Fiebrich, C.A.; Morgan, C.R.; McCombs, A.G.; Hall, P.K., Jr.; McPherson, R.A. Quality assurance procedures for mesoscale meteorological data. *J. Atmos. Ocean. Technol.* **2010**, *27*, 1565–1582. [CrossRef]

54. Venables, W.N.; Smith, D.M.; Team, R.C. An introduction to R-Notes on R: A programming environment for data analysis and graphics. 2018. Available online: https://cran.r-project.org/doc/manuals/r-release/R-intro.pdf (accessed on 15 March 2019)

55. SeaHawk. *SeaHawk-Desert Power Savonius Turbine*; Desert Power: Palm Desert, CA, USA, 2017.

56. Rabanal, A.; Ulazia, A.; Ibarra-Berastegi, G.; Sáenz, J.; Elosegui, U. MIDAS: A Benchmarking Multi-Criteria Method for the Identification of Defective Anemometers in Wind Farms. *Energies* **2018**, *12*, 28. [CrossRef]

Article

Desiccant-Assisted Air Conditioning System Relying on Solar and Geothermal Energy during Summer and Winter [†]

Peter Niemann *, Finn Richter, Arne Speerforck and Gerhard Schmitz

Institute of Engineering Thermodynamics, Hamburg University of Technology, 21073 Hamburg, Germany
* Correspondence: peter.niemann@tuhh.de; Tel.: +49-(0)40-42878-3273
† This paper is an extended version of our paper published in the 2018 International Sustainable Energy Conference (ISEC), Congress Graz, Austria, 3–5 October 2018.

Received: 19 July 2019; Accepted: 15 August 2019; Published: 19 August 2019

Abstract: At Hamburg University of Technology the combination of an open cycle desiccant-assisted air conditioning system and a geothermal system is investigated in the framework of different research projects for several years. The objective of this study is to investigate the energy efficiency of the overall system and to evaluate the geothermal system during summer and winter mode, based on data measured for a temperate climate region. Monitoring results of the performance for dehumidification and remoistening of supply air are presented. Furthermore, the investigated system is compared to reference air conditioning processes. During summer mode, an average dehumidification efficiency of 1.15 is achieved. The electrical energy savings compared to a conventional reference system sum up to 50% for the investigated cooling period. System operation during winter shows an average moisture recovery efficiency of 0.75. The electrical energy demand for air humidification is reduced by 50% compared to a system with electric isothermal air humidification. The geothermal system is operated efficiently throughout the year for cooling and heating application. Besides the energetic system evaluation, measured data regarding the soil temperature and thermal comfort are presented.

Keywords: air conditioning; borehole heat exchanger; desiccant dehumidification; enthalpy recovery; heat pump; system evaluation; experimental

---

1. Introduction

Due to increasing sales numbers and resulting energy demand for air conditioning as well as related CO_2 emissions worldwide, energy efficient and more environmental friendly air conditioning is required [1]. According to the International Energy Agency (IEA), an increase of more than five billion air conditioning systems between the years of 2016 and 2050 is estimated for the commercial and residential stock [2]. That means more than a doubling of the currently installed units. Income growth in the developing and emerging countries, climate change and increased building energy standards cause this development for air conditioning [3,4]. Currently, air conditioning is responsible for around 20% of buildings' electricity demand from a worldwide perspective [2]. Furthermore, heating, ventilation and air conditioning (HVAC) systems require the largest share of energy used in buildings. Thus, enhancing performance of conventional systems offers the opportunity to significantly reduce energy demand and related CO_2 emissions, respectively [5].

Air conditioning systems are often used to provide comfortable indoor air conditions in general. Removing latent and sensible loads from outside air is usually required during summer to provide the desired indoor air conditions. Especially moisture removal accounts for peak loads of conventional air conditioning systems since it requires cooling process air below dew point temperature. Cooling and dehumidification are coupled necessarily due to the process itself. Required cooling capacities are often provided by electrical driven vapor compression cycles. In contrast, removal of sensible and latent loads is separated within a desiccant assisted air conditioning process. A desiccant material is used to remove latent loads from the process air stream. Thus, required cooling capacities are reduced, especially at high outside air humidity ratios. Shallow geothermal energy can be utilized to remove sensible loads from the process air stream. Utilizing the soil for cooling, an equalized energy balance of the soil is essential regarding long-term efficiency of the geothermal system. This can be improved by using a ground-coupled heat pump for heat supply during winter.

1.1. Desiccant Assisted Air Conditioning

Valkiloroaya et al. [6] presented an overview of different strategies and technologies to reduce energy demands related to air conditioning in general. Desiccant-assisted air conditioning has been found to be a promising alternative to conventional air conditioning processes relying on a vapor compression chiller in terms of reducing the electricity demand for air conditioning. From a global perspective, air conditioning systems are primarily used during summer operation, providing cooled and dehumidified air. Thus, a lot of different studies dealing with the improvement and evaluation of desiccant materials and different system configurations for dehumidification mode. Several studies provide overviews of different concepts for desiccant assisted air conditioning systems with both solid and liquid desiccant material [7–11]. Within the field of systems relying on a solid desiccant material, a considerable amount of studies investigate design and performance of desiccant wheels [12–14]. Desiccant assisted hybrid systems are known as system configuration relying on an open sorption process and closed-loop cooling circuit. Several studies have been undertaken to investigate energetic advantages of hybrid systems for different locations [15–21]. To further reduce the electrical energy demand related to air conditioning, shallow geothermal energy is shown as promising alternative and renewable heat sink [22–25].

With respect to full year operation, final energy demand for space heating is currently still higher than final energy demand for space cooling applications from a global perspective [4]. But even though winter mode is obviously an essential part of full year operation, especially for heating dominated regions, winter mode as well as full year operation of desiccant assisted air conditioning systems are addressed only in few studies. Beccali et al. [26] investigated a hybrid system during summer and winter operation experimentally for the climate conditions of southern Italy. The presented system is relying on solar thermal heat supply with additional gas boiler backup system; a compression chiller is utilized for cooling. During summer operation, a reduction in primary energy demand of nearly 50% was achieved compared to a conventional reference system. Required information about system performance regarding moisture control in winter mode are not provided. De Antonellis et al. [27] investigated experimentally and numerically humidification of outside air using a desiccant wheel with silica gel coating for Mediterranean winter conditions. The authors investigated energy demand and occupants' discomfort for the considered system configuration and highlight the dependence of air humidification performance and required regeneration air temperature. To further evaluate the system's performance, a comparison with conventional humidification technologies is presented from an energetic point of view. Simulation results show reduced primary energy demand for air humidification using the proposed system compared to reference systems with adiabatic and electrical steam humidifiers for different working conditions. Compared to reference systems with steam to steam humidifier primary energy demand of the proposed system was increased for the considered boundary conditions. Kawamoto et al. [28] investigated the combination of a desiccant-assisted system and a heat pump that is used for heat supply on the regeneration air side experimentally in Japan. La et al. [29] examined a system configuration with solar thermal heat supply and one-rotor two-stage desiccant wheel for winter in Shanghai experimentally and numerically. The proposed system uses extract air from the conditioned space to humidify supply air preheated with solar thermal energy. The study shows significant increase in thermal comfort due to air humidification. Furthermore, the authors draw attention to the space requirements for solar collectors to improve thermal comfort. Full year operation of a desiccant- assisted evaporative system in Austria was investigated experimentally by Preisler and Brychta [30]. The investigated system achieved a reduction in primary energy demand of 60% in comparison to a reference system relying on a vapor compression chiller, considering full year operation. The authors outline high energy saving potentials of the investigated system, whereas details about the humidification process and boundary conditions of system comparison are not provided.

1.2. Air Dehumidification and Moisture Recovery

Regarding desiccant wheel performance in dehumidification and enthalpy recovery mode, Zhang and Niu [31] investigated different desiccant wheels numerically by means of a two-dimensional heat and mass transfer model. From their simulation results the authors conclude that the optimal rotational speed of a wheel used for dehumidification is much lower than optimal rotational speed of a wheel utilized for enthalpy recovery.

Increasing the moisture level of supply air is a sensitive but often little noticed comfort aspect during winter. Dry indoor air conditions can adversely affect occupants' comfort, especially in modern buildings relying on mechanical ventilation without additional humidification systems during winter. Conventional air conditioning systems require additional components to achieve sufficient supply air humidity ratios. This is an advantage of desiccant assisted systems, because moisture recovery by means of the existing hygroscopic material is possible. A further hygienic advantage of desiccant assisted moisture recovery against conventional air conditioning relying on adiabatic or isothermal air humidification is the fact that no liquid or vaporous water is sprayed into the process air stream. Thus, emission of bacteria caused by air humidifiers as for example described by Strindehag and Josefsson [32] is avoided.

1.3. Previous and Ongoing Investigations of the Considered System

To the best of the authors' knowledge there is no study investigating summer and winter operation of an air conditioning system relying on desiccant assisted dehumidification, enthalpy recovery and a ground-coupled heat pump for heating dominated climate conditions. In [33,34] the considered system is evaluated during summer operation mode, using a borehole heat exchanger (BHE) for cooling. The system is verified to be promising against conventional air conditioning systems in temperate climate regions. Furthermore, Speerforck et al. [35] proved applicability of the proposed system at different investigated locations by means of a dynamic system model using modeling language Modelica®. The authors investigated summer operation, whereas winter mode is not observed.

Within this study the geothermal and desiccant assisted system is investigated experimentally during summer and winter operation to show system performance throughout the year. Moisture control is achieved by a desiccant wheel (summer) or enthalpy wheel (winter) to improve indoor air conditions; sensible cooling loads are primarily covered by cooling ceilings, whereas heating loads are primarily covered by underfloor heating, respectively. In combination with a geothermal system, temperature levels of cooling ceilings and underfloor heating enable efficient operation of shallow geothermal energy in combination with a ground-coupled heat pump (GCHP) during winter. Regarding an equalized energy balance, utilizing the soil for cooling and heating is essential.

This study is structured into three parts. First, a short description of the investigated system, operation modes and data acquisition is provided. Afterwards, the performance and limitations of the investigated system are analyzed. Especially, performance of the air handling unit and the geothermal system are considered. The effects on indoor air conditions in terms of thermal comfort are investigated in detail. Additionally, the system is compared to different reference systems regarding electrical and thermal energy demands. Finally, the main findings are summarized and future research work is addressed. This study is an extension of Niemann et al. [36], providing a previous experimental analysis on summer and winter operation of the investigated system.

2. Materials and Methods

The investigated test facility is located on the campus of Hamburg University of Technology. Figure 1 shows some impressions of the installation and its relevant components. In total, the test facility consists of eight 20 ft. containers. The four containers on the lower floor contain the air handling unit and further technical installations. An office and conference room with a net floor space of 56 m^2 is located in the four upper containers. This area is used as reference room for the air conditioning system. System operation is investigated throughout the year.

Figure 1. Test facility, air handling unit and heat pump with parts of the header system.

2.1. Air Conditioning System

As shown in Figure 2, system layout of the installation can be divided basically into three subsystems in form of the reference room, the air handling unit and the hot and cold water circuit. The air handling unit is designed as hybrid system combining an open desiccant assisted air handling process with closed-loop heating and cooling circuits similar to the system presented by in [33,36].

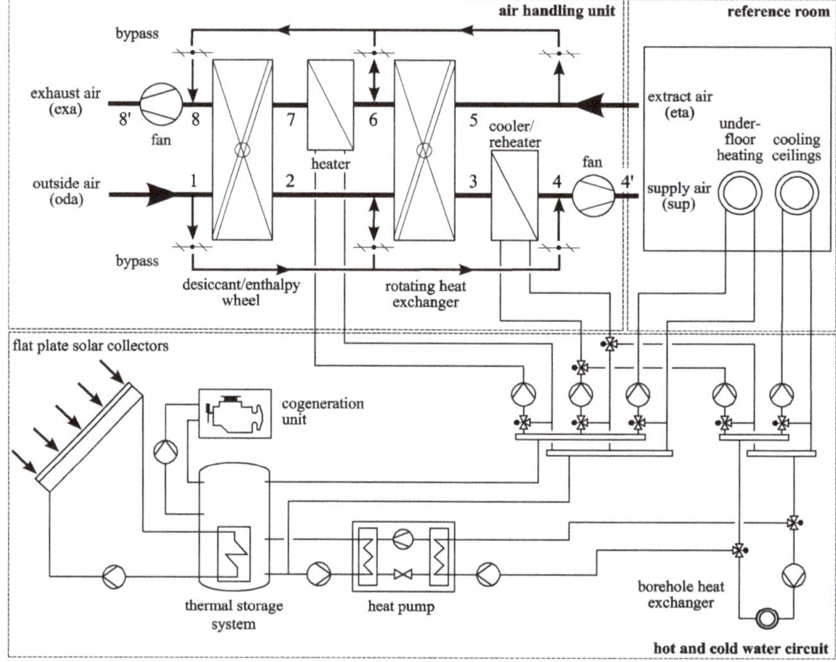

Figure 2. System layout of the test facility as used during summer and winter operation.

A brief description of system operation in summer and winter mode is given for the sake of completeness. Summer and winter operation are considered separately according to [36]. Considering dehumidification mode during summer, outside air (oda) is dehumidified within a desiccant wheel (1→2) and precooled by a sensible rotating heat exchanger (2→3). Water vapor is accumulated at the hygroscopic coating of the desiccant wheel (DW); lithium chloride (LiCl) is used as desiccant. Afterwards, process air is finally cooled or heated to the desired supply air (sup) temperature within a sensible water to air heat exchanger (3→4). Extract air (eta) from the reference room is preheated (5→6) by the heat recovery wheel (HRW) and further heated to the required regeneration air temperature within another sensible water to air heat exchanger (6→7). Finally, eta is used to regenerate the desiccant material (7→8), before it is emitted to the environment in form of exhaust air (exa). To achieve efficient operation, different components of the air handling unit can be bypassed as shown in Figure 2. Thus, electricity demand of the fans is reduced for demand-oriented air conditioning.

Regarding winter operation, the desiccant wheel is operated as enthalpy wheel at higher rotational speed for coupled heat and mass transfer (1→2) relying on passive air humidification. Oda is remoistened and reheated within this process using the eta stream. If oda humidity is within comfort limits regarding humidity ratio, it is preheated using the regenerative heat exchanger (2→3); the desiccant wheel is bypassed in this case. Otherwise, the heat recovery wheel is not utilized. The reheater (3→4) is used to adjust process air to the desired sup temperature. Eta is either used for sensible heat recovery (5→6) or coupled heat and moisture recovery (7→8). The heater (6→7) is not operated in winter operation mode.

Both wheels have a diameter of 0.6 m. The reference room is connected to the air handling unit on the supply and extract air side for air exchange. Furthermore, to cover sensible heat and cooling loads directly, it is equipped with underfloor heating and cooling ceilings.

Desiccant assisted air conditioning enables the integration of shallow geothermal energy for cooling in summer due to the fact that the required temperature level for cooling applications is above dew point temperature at any time. Due to the capacity of the soil, a cold water storage is not integrated into the cold water circuit with respect to summer operation. Utilizing the geothermal system during full year operation as heat sink and heat source is essential for the reason of improving the annual energy balance of the soil as well as for the reason of maximizing the use of renewable energies around the year. Thus, heat supply during winter is primarily relying on a ground-coupled heat pump system ($\dot{Q}_{GCHP,nom} = 5.1$ kW$_{th}$ at BW5/W30). Solar thermal energy is utilized as primary heat source during summer mode ($A_{STU} = 20$ m^2). A small-scale gas driven cogeneration (CHP) unit ($\dot{Q}_{CHP,nom} = 12.5$ kW$_{th}$, $P_{CHP,nom} = 5$ kW$_{el}$) is used as backup system and to cover peak loads throughout the year. Integrating a stratified thermal storage system ($V = 1$ m^3) into the hot water circuit enables heat supply and heat demand to be decoupled temporally.

2.2. Geothermal System

The geothermal system is built of a single double U-tube borehole heat exchanger (BHE), utilized as heat sink of the cold water circuit (summer) and heat source to supply the GCHP system (winter) as shown in Figure 2. With respect to the geothermal system itself, Figure 3 shows the structure of the soil at the drilling location and the design of the BHE. The thermistor string of an additional reference BHE (Ref) of the same type is used to analyze the impact of utilizing geothermal energy on the surrounding soil.

The soil primarily consists of micaceous clay. Nevertheless, for the first 18 m below ground surface fine, medium and coarse sands are present. A layer of till and silt is located underneath. There are no relevant ground water flows at the drilling location in general. Thermal conductivity of the grouting material is $\lambda = 2$ W $\cdot$m^{-1} $\cdot$K^{-1}.

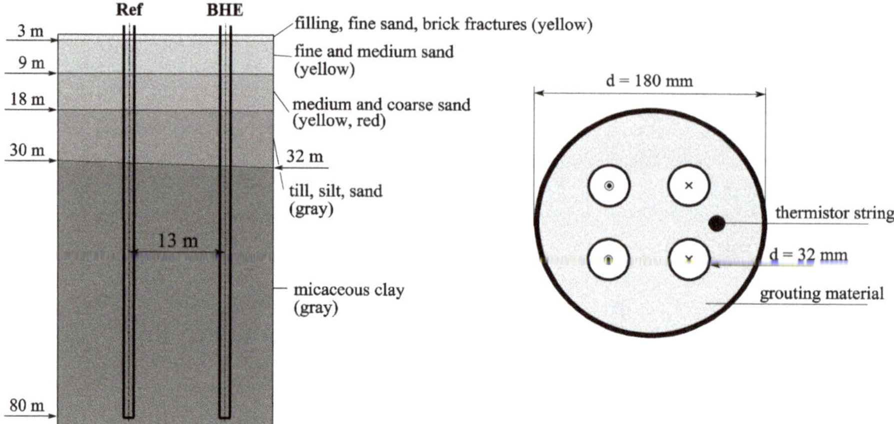

Figure 3. Structure of the soil and design of the borehole heat exchangers.

2.3. Data Acquisition

To characterize the status of working fluids within the overall process, all relevant parameters are measured and recorded. Table 1 provides an overview of the measurement characteristics for the entire process. Air states are labeled according to Figure 2.

Table 1. Data acquisition concept for the entire system separated by subsystems.

Subsystem	Measured Value	Comment
AHU	Air temperature and relative humidity Pressured drop Air volume flow	Inlet and outlet of each component Across each component At positions 4 and 8 for sup and eta
Hydraulic circuits	Fluid volume flow Fluid temperature	Inlet or outlet of each circuit Inlet and outlet of each circuit
BHE	Soil temperature	Thermistor string embedded in the borehole; temperature measurement in depths of 10, 15, 20, 40, 60 and 80 m
Reference room	Thermal comfort	Temperature and humidity ratio within the conditioned space

In order to take inhomogeneity within the air streams into account, flow averaging is applied according to Slayzak and Ryan [37]. Electricity demand of each component according to Figure 2 is measured separately. Measurement devices in use and related uncertainties are listed in Table 2. All measured data are recorded every minute. A data acquisition system in connection with controlling software is used to control and regulate the entire system.

Due to drifting effects of capacitive humidity sensors (typical 1% rh per year), the resilience of derived quantities, energy and mass balances gets more and more limited as time goes by. Thus, these sensors are recalibrated once a year. The calibration method is relying on 36 set point combinations of temperature and relative humidity in the range of relevant temperatures and humidity ratios.

Table 2. Measurement devices and related uncertainties.

Measured Value	Sensor Type or Principle	Measurement Uncertainty
Air/water temperature	Pt 100 (accuracy class W 0.1)	$\pm\, 1/3 \cdot (0.3 + 0.005 \cdot \vartheta)$ K
Soil temperature	Thermistor string	$\pm\, 0.5$ K
Relative humidity	Capacitive humidity sensor	$\pm\, 2\%$ rh for $10\ldots90\%$ rh
Volume flow (air)	Differential pressure	$\pm\, 10\%$ of reading
Volume flow (water)	Electromagnetic flow meter	$\pm\, 0.5\% \pm 1$ mm $\cdot$s^{-1} of reading
Pressure difference	Ceramic fulcrum lever technology	$\pm\, 2\%$ of full scale (range: $0\ldots300$ Pa or $0\ldots1000$ Pa)
Electric power	AC energy meter	$\pm\, 2\%$ of reading

3. Results and Discussion

The results presented in this study are based on measured data of the cooling period from June till September in 2016 and the following heating period from January till March in 2017. During the considered periods the test facility was operated from 7 am to 10 pm every day of the week. Transition periods in spring and fall are not considered in this study, because these periods are not suitable to analyze strengths and weaknesses of the system as a reason of the climate conditions in northern Germany. Volume flow of supply air was controlled to be constant in the range of (950 ± 95) m^3 $\cdot$h^{-1}; mass flow rates of supply and extract air were controlled to be equal. Set point of sup water content is 8 g$_\text{w}$ $\cdot$kg$_\text{air}^{-1}$ for dehumidification mode. The following evaluation is subdivided into four parts. First, the system is evaluated regarding relevant performance parameters of summer and winter operation and the performance of the geothermal system is evaluated. Afterwards, thermal comfort within the air conditioned space is analyzed. Finally, the investigated system is compared to different reference systems focusing electrical and thermal energy demands. System performance is evaluated separately for summer and winter operation in general for this study in order to show strengths and weaknesses for each operation mode.

3.1. Performance Evaluation of the Air Conditioning System

The following evaluation of system performance is based on measured data during the investigated periods. First, electrical power demand is considered. Electrical power demand of the entire system is in the range of 770–900 W$_\text{el}$ during summer operation. The fans account for about 81–95% of this power demand, whereas the remaining part is divided equally between other auxiliary energies (e.g., drive of the wheels, circulation pumps). The electrical power demand of the GCHP has to be considered additionally during winter operation ($P_\text{GCHP} = 887 - 1388$ W$_\text{el}$). Indexing within the following equations is according to Figure 2.

To evaluate the air handling unit for the considered periods, electrical and thermal COP values are used. These performance indicators are defined according to [33]:

$$\text{COP}_\text{el,AHU,su} = \frac{\dot{m}_\text{sup} \cdot (h_1 - h_{4'})}{P_\text{el,AHU}} \qquad \text{COP}_\text{el,AHU,wi} = \frac{\dot{m}_\text{sup} \cdot (h_{4'} - h_1)}{P_\text{el,AHU}} \tag{1}$$

$$\text{COP}_\text{th,AHU,su} = \frac{\dot{m}_\text{sup} \cdot (h_1 - h_{4'})}{\dot{Q}_\text{th,AHU,su}} \qquad \text{COP}_\text{th,AHU,wi} = \frac{\dot{m}_\text{sup} \cdot (h_1 - h_{4'})}{\dot{Q}_\text{th,AHU,wi}} \tag{2}$$

$$\dot{Q}_\text{th,AHU,su} = \dot{m}_\text{w,AH} \cdot (h_\text{w,in,AH} - h_\text{w,out,AH}) \qquad \dot{Q}_\text{th,AHU,wi} = \dot{m}_\text{w,RH} \cdot (h_\text{w,in,RH} - h_\text{w,out,RH}) \tag{3}$$

All figures shown in the following rely on steady-state operation conditions. Measured data were selected for steady state operation 15 minutes after the last changes made by system control. The electrical and thermal COP of the air handling unit during summer operation at dehumidification mode are shown in Figure 4 in dependence of regeneration air temperature. A strong dependence between COP$_\text{el,AHU,su}$ and regeneration air temperature is visible from the plot in Figure 4a. Due to the fact that the electrical energy demand of the AHU is nearly independent of the outside air conditions, $COP_{el,AHU,su}$ is increased with increasing regeneration air temperature. Increase in regeneration air temperature is a result of increasing water content and/or temperature of outside air within dehumidification mode. With respect to Figure 4b, desired supply air temperature cannot be maintained at high outside air temperature and water content, causing an increase in supply air enthalpy, respectively. This causes flattening of the curve for $COP_{el,AHU,su}$ with increasing regeneration air temperature. Due to its definition, negative values of $COP_{el,AHU,su}$ occur at low outside air temperatures when dehumidification of supply air is necessary whereas cooling is not required.

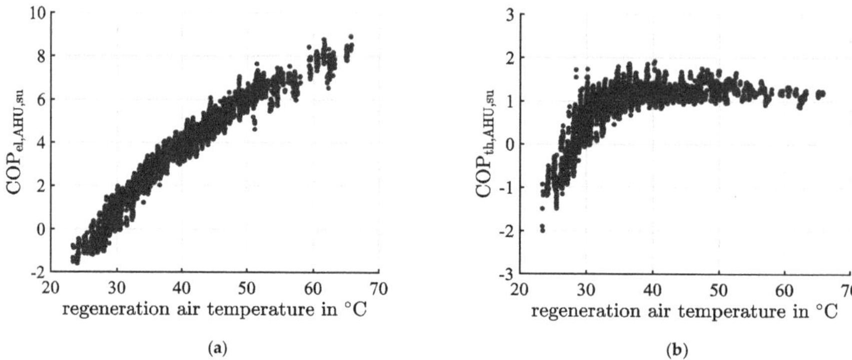

(a)　　　　　　　　　　　　　(b)

Figure 4. (**a**) Electrical COP of the air handling unit during summer operation at dehumidification mode in dependence of regeneration air temperature; (**b**) thermal COP of the air handling unit during summer operation at dehumidification mode in dependence of regeneration air temperature.

At values of regeneration air temperature below 35 °C, the increase of AHU thermal COP with increasing regeneration air temperature is much steeper compared to the slope of $COP_{th,AHU,su}$ at higher regeneration air temperature. This characteristic results from the mathematical definition of $COP_{th,AHU,su}$ as presented in Equation (2). Negative values of $COP_{th,AHU,su}$ occur at low regeneration air temperature when dehumidification is still necessary and reheating of supply air is required at the same time. For higher regeneration air temperatures above 35 °C, thermal COP keeps nearly constant at $COP_{th,AHU,su} \geq 1$ with decreasing fluctuations for increasing regeneration air temperature.

For the investigated heating period, performance indicators in form of $COP_{el,AHU,wi}$ and $COP_{th,AHU,wi}$ are shown in Figure 5. To evaluate the performance of the air handling unit during winter mode, the characteristics of $COP_{el,AHU,wi}$ and $COP_{th,AHU,wi}$ depending on outside air temperature are presented for EW mode and HRW mode.

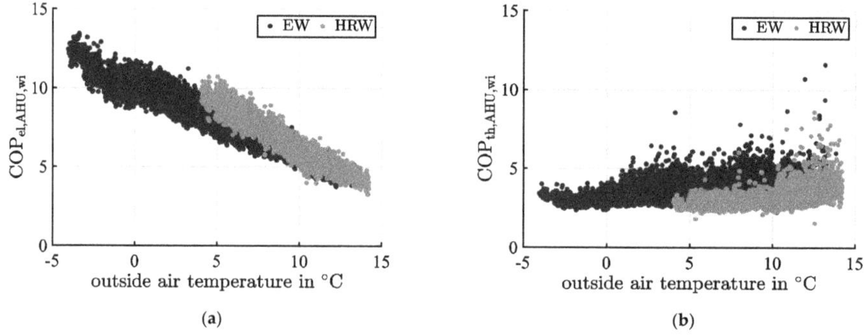

(a)　　　　　　　　　　　　　(b)

Figure 5. (**a**) Electrical COP of the air handling unit during winter operation in dependence of outside air temperature; (**b**) thermal COP of the air handling unit during winter operation in dependence of outside air temperature.

As shown in Figure 5a, the electrical COP of the air handling unit is decreasing with nearly constant gradient for increasing outside air temperature in both operation modes. With respect to Equation (1), this is a result of decreasing nominator with increasing outside air temperature, whereas the denominator keeps constant in good approximation. Generally, water content of outside air is sufficient regarding indoor air comfort limits at higher outside air temperature level. Thus, HRW mode is just occurring at outside air temperatures above 4 °C. Due to similar electrical energy demand of the wheels as well as similar pressure drop across these components, the resulting gradients of $COP_{el,AHU,wi}$ are similar. The dependence of thermal COP on outside air temperature for the air handling unit during winter operation is shown in Figure 5b. In spite of increasing fluctuations with increasing outside air temperature, $COP_{th,AHU,wi}$ is independent of outside air temperature and not dropping below $COP_{th,AHU,wi} = 2$. This is a result of its definition, as presented in Equation (2), with similar characteristic of nominator and denominator for the underlying boundary conditions.

To further evaluate desiccant assisted dehumidification and enthalpy recovery in detail, mode-specific key figures have to be defined. In order to take the fact into account that the desiccant wheel is used for active dehumidification with a regeneration air heater, dehumidification efficiency is defined as follows:

$$DCOP = \frac{\dot{m}_{sup} \cdot (x_1 - x_{4'}) \cdot r_0}{\dot{Q}_{th,AHU,su}}, \tag{4}$$

with $\dot{Q}_{th,AHU,su}$ according to Equation (3) and is therefore equivalent to the definition of latent COP. An average dehumidification efficiency of $\overline{DCOP} = 1.15$ was achieved for the considered cooling period. This result indicates that more latent thermal power was absorbed within the desiccant wheel compared to the required thermal power to run the regeneration air heater. In general, regenerative heat exchange within the air handling unit improves dehumidification efficiency by preheating extract air for regenerating the desiccant wheel. Due to the fact that the wheel is used for passive enthalpy recovery during winter, moisture recovery efficiency is expressed by:

$$\Psi = \frac{x_2 - x_1}{x_7 - x_1}. \tag{5}$$

An average moisture recovery efficiency of $\overline{\Psi} = 0.75$ was achieved for the enthalpy wheel during the investigated winter period. Thus, an increase of 1.1 $g_w\ kg_{air}^{-1}$ was achieved for sup humidity ratio on average. Maximum values of moisture recovery were close to 2.3 $g_w\ kg_{air}^{-1}$ for the underlying boundary conditions.

3.2. Performance Evaluation of the Geothermal System

Performance evaluation of the overall geothermal system is structured into three parts. First, soil temperature and soil energy balance are considered. Afterwards, energy transfer at the BHE is investigated and finally, GCHP performance is evaluated.

Temperature profiles of BHE and the considered reference BHE are shown in Figure 6a for the period of around one year including summer and winter operation. The soil temperature 15 m below ground surface was found to be independent of seasonal related temperature fluctuations during previous investigations. Thus, the plots result from temperature averaging below 15 m for both, BHE and reference BHE. Depending on the season and operation mode of the air conditioning system, the the soil around the BHE is significantly influenced with dynamic temperature profile during cooling and heating mode. Cooling peak loads occurring in summer operation lead to maximum soil temperatures above 18 °C. This temperature level is crucial with respect to keep the desired indoor air temperature level below 25.5 °C. Regardless of such peak loads that only occurred at a few days of the considered cooling period, the soil temperature was kept within a sufficient temperature range in terms of cooling purposes. During winter operation, the soil temperature is less fluctuating compared to its use as heat sink. The lowest soil temperature was 4.5 °C that occurred during the coldest period in the beginning of February. Using the soil for heating, the soil temperature decrease is crucial to operate the GCHP system efficiently. This dependence is further analyzed later on in this subsection. Average temperature level of the undisturbed soil at the reference BHE was at 9.8 °C. Occurring temperature fluctuations were within the corresponding uncertainty of temperature measurement.

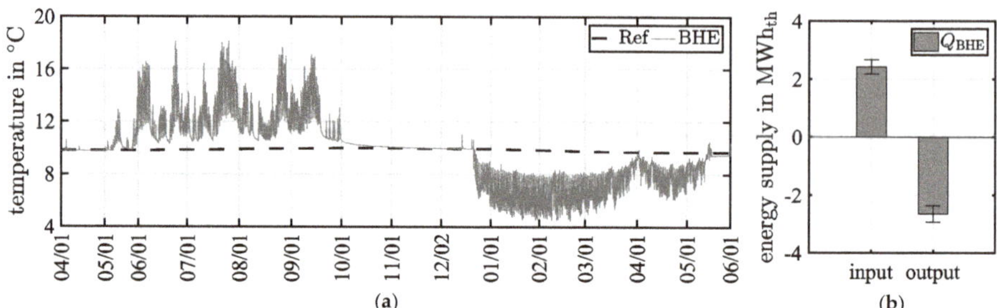

Figure 6. (**a**) Temperature profile of the grouting material for both borehole heat exchangers; (**b**) soil energy balance based on thermal energy input and output.

Balancing input and output of thermal energy during summer and winter operation and natural regeneration of the soil, an equalized energy balance of the soil was achieved. Input and output of thermal energy at the BHE are balanced with a remaining annual difference of 0.22 MWh$_\text{th}$, 9% respectively, as shown in Figure 6b. This difference is within measurement uncertainty of the corresponding energy values. With respect to these results an efficient long-term operation of the geothermal system can be predicted. Nevertheless, an ongoing long-term monitoring of the geothermal system is essential, especially when large scale geothermal systems with several BHE influencing each other are considered.

In terms of further investigating energetic performance of the geothermal system, month and period specific performance indicators are presented in Figure 7. For both periods the amount of energy (Q_BHE) and heat flow ($\dot{Q}_\text{BHE}$) transferred at the BHE as well as resulting performance values are shown. Key figures as defined in [33] are used to evaluate BHE performance. With respect to the considered periods, Monthly Performance Factor (MPF) and Seasonal Performance Factor (SPF) are used as presented in Equation (6):

$$\text{MPF} = \frac{\int_\text{m} \left| \dot{Q}_\text{BHE} \right| d\tau}{\int_\text{m} P_\text{PU} \, d\tau} \qquad \text{SPF} = \frac{\int_\text{p} \left| \dot{Q}_\text{BHE} \right| d\tau}{\int_\text{p} P_\text{PU} \, d\tau} \tag{6}$$

The denominator includes the electrical energy demand of the BHE circulation pump. Decreasing MPF values occurred over each period as a result of changing temperature level of the soil surrounding the BHE by charging or discharging energy in form of heat. The amount of thermal energy, heat flow and resulting MPF show the same relationship for both periods with one exception. Even though the month of July shows the largest amount of thermal energy transferred to the soil, the highest MPF was achieved in June with $\text{MPF}_\text{su,max} = 170 \pm 17$, see Figure 7a. This was primarily caused by lower temperature increase of the soil during June. Evaluating the entire cooling period, a seasonal performance of $\text{SPF}_\text{su} = 153 \pm 15$ was achieved, indicating a high efficiency of the geothermal heat sink. The same holds true for the winter period with a resulting seasonal performance of $\text{SPF}_\text{wi} = 110 \pm 11$, even though the value is around 28% lower compared to summer mode. The reason for this difference is related to the required volume flow of heat transfer medium to supply the evaporator of the heat pump that is generally higher than volume flows of heat transfer medium for natural cooling in summer.

To further analyze GCHP performance, its electrical COP, see Equation (7), is investigated in more detail in terms of available and required temperature levels as shown in Figure 8:

$$\mathrm{COP_{GCHP}} = \frac{\left|\dot{Q}_h\right|}{P_{\mathrm{GCHP}} + P_{\mathrm{AUX}}} \tag{7}$$

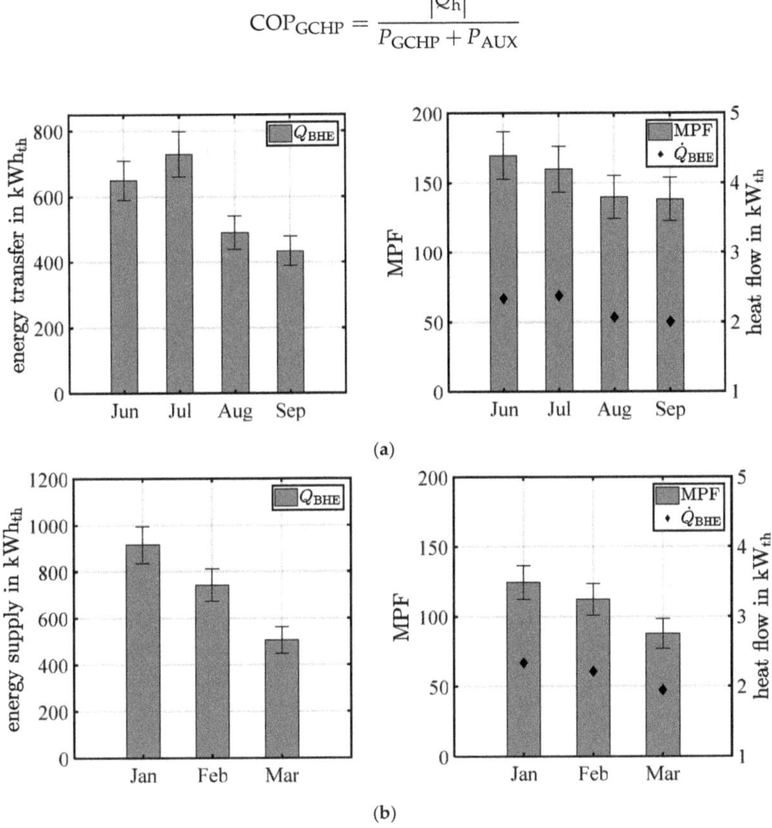

Figure 7. Thermal energy transfer at the BHE (**left**) and performance parameters of the BHE (**right**) for the investigated periods: (**a**) cooling period; (**b**) heating period.

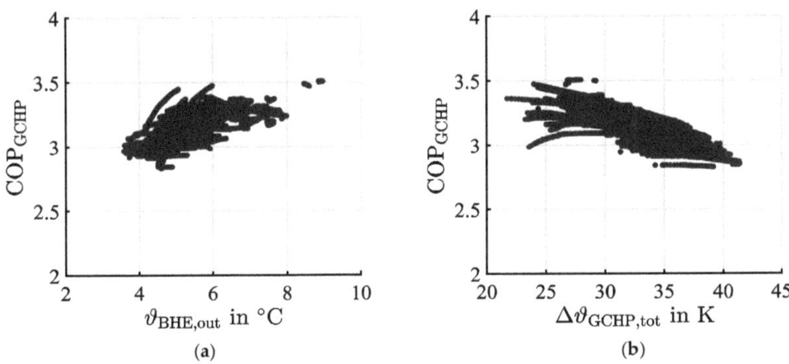

Figure 8. (**a**) GCHP electrical COP in dependence of BHE outlet temperature $\vartheta_{\mathrm{BHE,out}}$; (**b**) total GCHP temperature lift $\Delta\vartheta_{\mathrm{GCHP,tot}}$.

BHE outlet temperatures were mostly pooled in the range of $\vartheta_{\text{BHE,out}} = 3.8 - 8\,°C$ during steady state GCHP operation with a range of $\text{COP}_{\text{GCHP}} = 2.9 - 3.5$, as shown in Figure 8a. A trend of increasing COP_{GCHP} with increasing BHE outlet temperature is visible. This is an effect of lower required temperature lift within GCHP process that comes along with reduced power level required to run the compressor. Taking the required temperature lift as a further indicator of GCHP load into account, a slight dependence on GCHP performance can be deduced from Figure 8b. The temperature lift $\Delta\vartheta_{\text{GCHP,tot}}$ is defined as temperature difference between condenser outlet and evaporator inlet. Values above $\Delta\vartheta_{\text{GCHP,tot}} = 32\,K$ that were required to supply UHS caused GCHP performance lower than 3 with decreasing trend curve. Taking the overall heating period into account, SPF of the GCHP system can be determined equivalent to Equation (6) by integrating thermal and electrical powers from Equation (7):

$$\text{SPF}_{\text{GCHP}} = \frac{\int_{\text{p}} \left|\dot{Q}_{\text{h}}\right| \, d\tau}{\int_{\text{p}} \left(P_{\text{GCHP}} + P_{\text{AUX}}\right) d\tau} \tag{8}$$

For the considered heating period $\text{SPF}_{\text{GCHP}} = 3$ was achieved. Compared to GCHP systems state of the art with $\text{SPF}_{\text{GCHP}} = 4.0 - 4.5$, performance of the investigated system relying on a reciprocating compressor shows potential for improvement. Nevertheless, this system is robust against fluctuating temperature levels.

3.3. Evaluation of Thermal Comfort

The quality of indoor air conditions is a result of the performance and operation strategy of the overall system. To provide an overview of comfort conditions for the investigated periods, outside and room air conditions during system operation are shown in Figures 9a and 10a with simplified comfort areas according to DIN EN 15251 [38]. Comfort areas of category I and II are defined to ensure less than 6% (cat. I) and 10% (cat. II) of occupants being dissatisfied with the present indoor air conditions. In order to exclude start-up effects, the first hour of system operation is not considered.

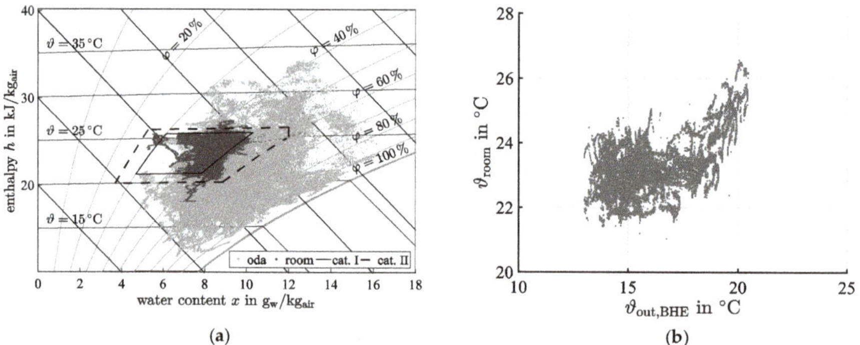

Figure 9. (**a**) Outside and room air conditions during system operation for the investigated cooling period; (**b**) dependence of room air temperature on BHE outlet temperature.

For the investigated cooling period, indoor air conditions according to cat. I were maintained for 55% and cat. II was maintained for over 96% of operation time, respectively. The reasons for the remaining violations were different for cat. I and cat. II. Cat. I was violated primarily due to too high indoor air humidity, whereas too high indoor air temperatures caused violations of cat. II.

In order to further analyze the reasons for overheating, Figure 9b shows the dependence of room air temperature ϑ_{room} on water outlet temperature $\vartheta_{\text{out,BHE}}$ of the BHE. Outlet temperature of the BHE is similar to the cooling ceilings' inlet temperature. With good approximation, a linear increase of indoor air temperature with increasing BHE outlet temperature above $18\,°C$ is visible. Maximum room air temperatures were right above 26 °C at maximum BHE outlet temperature of $\vartheta_{\text{out,BHE,max}} = 20.5\,°\text{C}$. This is an effect of peak loads that could not be covered by the geothermal heat sink due to its limited capacity and little controllable thermal power output.

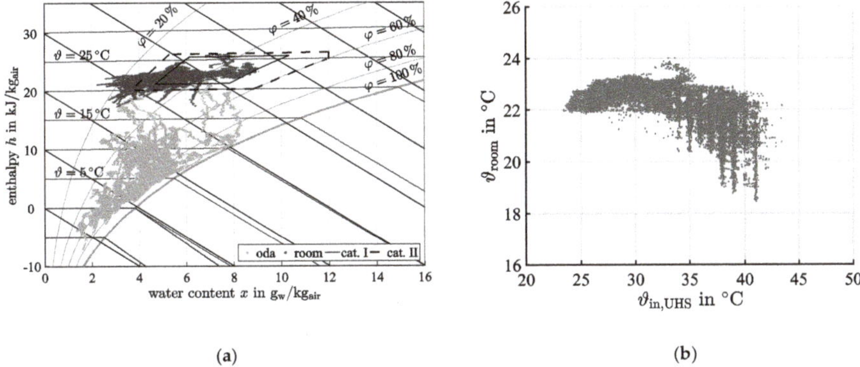

(a) (b)

Figure 10. (a) Outside and room air conditions during system operation for the investigated heating period; (b) dependence of room air temperature on underfloor heating inlet temperature.

According to the German Meteorological Service, the investigated heating period can be classified as moderate. The average oda temperature in January was 1.67 °C at an average water content of 3.3 $g_{\text{w}} \cdot \text{kg}_{\text{air}}^{-1}$. The following months were characterized by higher oda temperature and water content. As shown in Figure 10a, nearly 100% of oda conditions were outside comfort area according to cat. I and II. Days with oda conditions within the desired comfort area just occurred during springtime period at the end of March.

Around 67% of indoor air conditions satisfied the requirements according to cat. I. Cat. II was maintained for 75% of system operation time, respectively. The remaining violations were primarily caused by too low indoor air temperature. As a result of the control strategy for the underfloor heating system (UHS) and insufficient internal loads, 85% of indoor air temperatures below the desired temperature level occurred during the first four hours of daily system operation. A similar characteristic applied to indoor air water content. The level of sup and indoor air water content is mainly depending on oda water content if constant internal latent loads are provided. Thus, 75% of indoor air conditions characterized by insufficient level of water content occurred at oda water contents below 2.5 $g_{\text{w}} \cdot \text{kg}_{\text{air}}^{-1}$. Without humidification of supply air, over 60% of room air conditions would have been outside of cat. I. Over-humidification of room air did not occur during the entire heating period.

Figure 10b shows the dependence of indoor air temperature ϑ_{room} on UHS inlet temperature $\vartheta_{\text{in,UHS}}$. A trend of increasing fluctuation of indoor air temperature with increasing UHS inlet temperature is noticable, especially for $\vartheta_{\text{in,UHS}} > 32\,°\text{C}$. This is an effect of insufficient thermal power supply during the first hours after system start-up at low oda temperatures. Taking this initial situation, improvement of the control strategy for the UHS is required in terms of limiting periods at insufficient thermal comfort. The GCHP system was not operated between 10 pm and 7 am, while underfloor heating was provided 24 hours a day. Providing and storing sufficient amounts of thermal energy can be achieved by operating the GCHP system additionally during night. Resulting shorter regeneration periods of the soil have to be considered for the proposed operation strategy focusing on increase in thermal comfort.

Table 3 lists the relative shares of operation modes for both periods for the given outside air conditions and the implemented control strategies. The significant amount of DW or EW mode represents the importance of desiccant assisted air conditioning to provide a high level of indoor air conditions throughout the year.

Table 3. Relative shares of operation modes.

Operation Mode	Summer	Winter
DW (su)/EW (wi)	76%	70%
HRW	19%	30%
Simple air exchange	5%	<1%

3.4. System Comparison

The investigated system is compared to different reference systems in order to further evaluate its energetic performance. These reference systems are designed as mathematical simulation models relying on measured data of the investigated system (DW-GEO). Thus, the thermodynamic processes of the considered reference systems were modeled relying on simplified thermodynamic relations and required assumptions. Measurement data of the investigated system were used whenever similar air states are expected for the reference systems in comparison to DW-GEO. In order to ensure a fair system comparison, oda and sup conditions as well as sup mass flow rate and room air conditions were assumed to be equal with one exception that is explained later on. Air dehumidification as well as air humidification within the reference systems are considered according to the equivalent operation modes of the investigated system. In each operation mode, summer and winter, two different reference systems were designed. During summer operation, reference systems relying on an electrical powered vapor compression chiller are considered. Humidifying processes with adiabatic or isothermal air humidification are considered for the reference systems regarding winter operation. The individual characteristics of each reference system are summarized in Table 4.

Table 4. Characteristics of reference systems.

Parameter	DP-VC	DW-VC	AH-GEO	IH-GEO
Related period	Summer	Summer	Winter	Winter
Air dehumidification	Cooling below dew point tem-perature	Similar to DW-GEO	Not considered	Not considered
Air humidification	Not considered	Not considered	Electrical impeller humidifier (adiabatic)	Electrical steam humidifier (isothermal)
Supply air temperature	18 °C	22 °C	22 °C	22 °C
Heating power	Similar to DW-GEO, adjusted	Similar to DW-GEO	Similar to DW-GEO, adjusted	Similar to DW-GEO, adjusted
Cooling power	Vapor compres-sion chiller ($EER_{el} = 3$)	Vapor compres-sion chiller ($EER_{el} = 3.2$)	Not considered	Not considered

Supply air temperature of the conventional reference system (DP-VC) was set to 18 °C in order to ensure a fair system comparison. Changes in thermal load discharge by air and cooling ceilings was considered. The second reference system for the cooling period is a hybrid air conditioning system (DW-VC); the BHE is replaced by a vapor compression chiller, while the rest of the investigated system remains unchanged. The specific layout of corresponding air handling units for the reference systems DP-VC and AH-GEO/IH-GEO is shown in Figure 11. Highlighted components are specific for the designated reference systems, whereas the rest of the air handling unit remains the same for these systems.

Figure 12 shows the results of the system comparison for the considered periods. Legend entry "supply" includes electricity demands of the compression chiller (DP-VC, DW-VC) or rather the BHE circulation pump (DW-GEO) during summer operation, see Figure 12a. During winter operation, shown in Figure 12b, it includes the heat pump and corresponding auxiliary energies, respectively. Electricity demand of the AHU and circulation pumps of the hydraulic circuits are summarized to category "air treatment and distribution" for both periods. Electrical and thermal energy demands are related to the entire periods under consideration. All values of electrical and thermal energy demands shown in Figure 12 are also listed in Table 5 for more transparency.

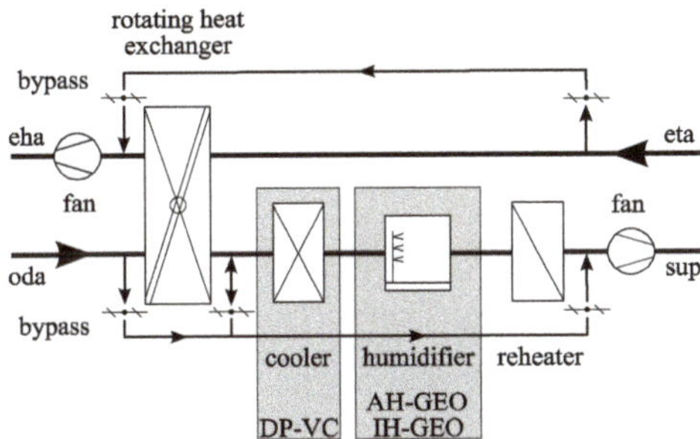

Figure 11. Specific layout of air handling units for different reference systems.

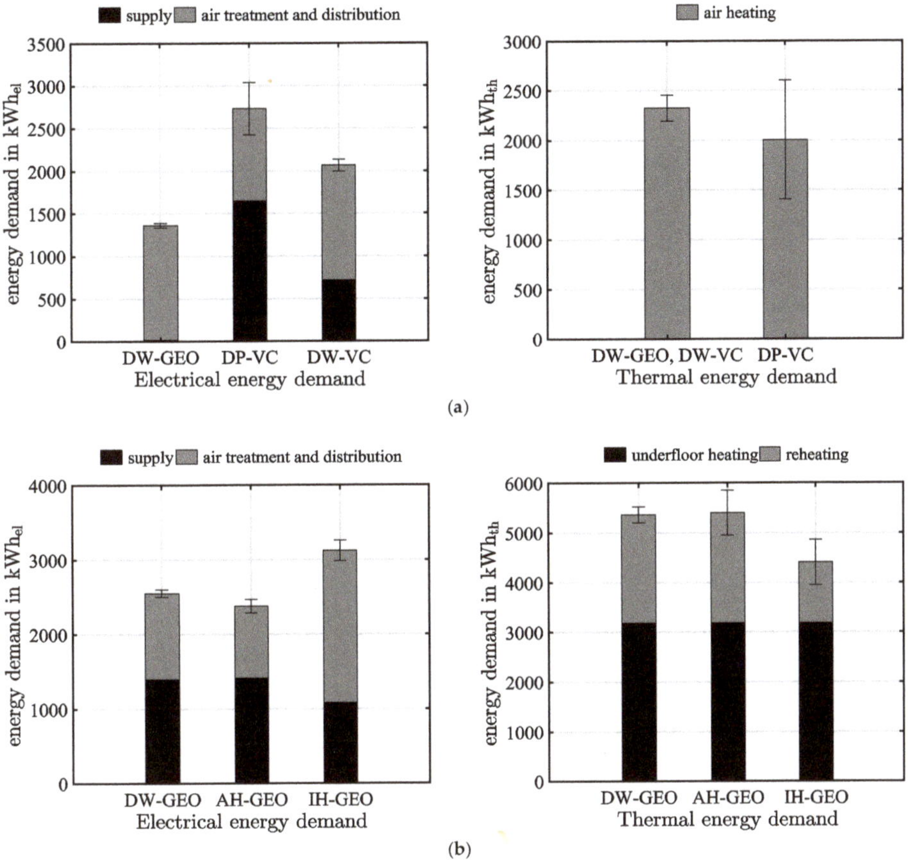

(**a**)

(**b**)

Figure 12. Total electrical and thermal energy demands of the investigated system (DW-GEO) and reference systems (summer: DP-VC, DW-VC; winter: AH-GEO, IH-GEO) for the investigated periods: (**a**) cooling period; (**b**) heating period.

Table 5. Electrical and thermal energy demands for system comparison.

Energy Demand	DW-GEO	DP-VC	DW-VC	AH-GEO	IH-GEO
Cooling period					
Supply (in kWh_{el})	15	1649	720	–	–
Air treatment and distribution (in kWh_{el})	1348	1082	1348	–	–
Air heating (in kWh_{th})	2326	2004	2326	–	–
Heating period					
Supply (in kWh_{el})	1401	–	–	1414	1082
Air treatment and distribution (in kWh_{el})	1153	–	–	964	2043
Underfloor heating (in kWh_{th})	3180	–	–	3180	3180
Reheating (kWh_{th})	2182	–	–	2223	1225

As shown in Figure 12a, the investigated system boasts significant reductions in electrical energy demand compared to the reference systems during summer operation. The savings sum up to 50% compared to the conventional system (DP-VC) and 34% compared to the hybrid system (DW-VC). These savings were primarily caused by the chiller unit (black column) that required 61% of the total electrical energy demand for the system DP-VC, whereas just 1% of the system DW-GEO was required for the BHE circulation pump. Due to the fact that only sensible cooling is required for the chiller unit of the hybrid system, its electricity demand was reduced by more than 50% compared to the conventional system. The overall AHU electricity demand of the system DP-VC was lower compared to the desiccant assisted systems (DW-GEO, DW-VC), because pressure drop across the desiccant wheel was saved for this system. Considering thermal energy demands, the conventional system shows the lowest thermal energy demand for reheating supply air. The required regeneration process within the desiccant assisted systems caused an increase of thermal energy demand by a factor of 1.2 compared to the system DP-VC. In total, the differences in thermal energy demand were not significantly high caused by the moderate summer period with a certain amount of oda conditions that required reheating but no or only moderate dehumidification. Nevertheless, the increased thermal energy demand of desiccant assisted air conditioning requires a convenient and low-cost heat source regarding primary energy demand of heat supply.

During winter operation, as shown in Figure 12b, the benefits of the investigated system (DW-GEO) are not as obvious as in summer mode. Thus, electrical and thermal energy demands have to be analyzed carefully. For the reference system with electric isothermal air humidification (IH-GEO) electricity demand for air treatment and distribution was increased by a factor of 1.8 compared to DW-GEO. This was significantly induced by the electrical steam humidifier that accounts for nearly 50% of the corresponding gray column. The corresponding electricity demand for AH-GEO was reduced by 16% compared to DW-GEO for the following reasons. First, the additional electricity demand for the impeller humidifier is low and second, the pressure drop associated with the enthalpy wheel is saved. This applies for IH-GEO, respectively. Due to the reasons mentioned above, the electricity demand to operate the GCHP system is almost equal for the systems DW-GEO and AH-GEO. For the reason of high temperature steam used to humidify supply air within the system IH-GEO, required GCHP power was lower in terms of adjusting sup stream to the desired sup temperature. In total, the electricity demand of DW-GEO was reduced by 18% compared to IH-GEO and it was increased by 7% compared to AH-GEO.

Thermal energy demand required for underfloor heating is assumed to be equal for the three systems. Thermal energy required for reheating sup to the desired sup temperature was about 44% higher for the DW-GEO and AH-GEO systems compared to the system with isothermal air humidification due to substitution of thermal energy by electricity as described above. Even though the differences are quite small, considering both, electrical and thermal energy demands, system comparison shows lowest total energy demand for AH-GEO. This results from the slight advantage in electricity demand of AH-GEO against the investigated system. Thus, energetic benefits of the investigated system were limited for the considered heating period. It is expected that increased benefits will be achieved for winter terms with lower oda temperature and water content. Nevertheless, moisture recovery using an enthalpy wheel is beneficial against the reference technologies from a hygienic point of view, especially for the use of LiCl desiccant material.

Summarizing, taking summer and winter operation into account, the investigated system boasts significant reductions in electricity demand, resulting from the electricity savings during summer operation. Additional thermal energy demand required for regeneration of the desiccant wheel is not significantly higher at the same time. The required temperature level up to 70 °C can commonly be easily provided by solar thermal energy. Nevertheless, providing thermal energy as favorable as possible is crucial considering the economic efficiency of the system. Another advantage of the investigated system is the reduced amount of mode specific equipment. Subsystems like the desiccant wheel and the geothermal system of the investigated system are used throughout the year, whereas this holds not true for the chiller and the air humidifier of the reference systems.

4. Conclusions

In this study experimental investigations of an air conditioning system during summer and winter mode relying on desiccant assisted dehumidification and enthalpy recovery and a geothermal system used for cooling and heating are presented. The investigated system is able to fulfill the requirements regarding comfortable indoor air conditions during the investigated periods. A desiccant wheel based on LiCl is used for air dehumidification and remoistening, achieving an average dehumidification efficiency of 1.15 and a moisture recovery efficiency of 0.75. Compared to reference processes the investigated system is beneficial during summer mode. Savings in electrical energy demand of up to 50% are achieved compared to a conventional air conditioning process. In contrast, a reference system relying on adiabatic air humidification shows slight benefits for the investigated heating period. Nevertheless, desiccant assisted humidification is advantageous compared to other humidification processes regarding hygienic aspects. If full year operation of an air conditioning system requires air dehumidification during summer it is essential to make use of the pre-existing desiccant material for remoistening of process air during winter. Additionally, the amount of mode specific equipment is reduced for desiccant assisted air conditioning with respect to full year operation. Required temperature levels of the working fluids enable the energetic use of a geothermal system for cooling and heating applications. The thermal energy balance of the soil is equalized throughout full year operation, enabling the soil to be used as heat source and heat sink for long-term periods. Thus, the amount of renewable energy sources is increased significantly to ensure environmental friendly air conditioning.

The research project includes detailed investigation of thermal comfort and economic assessment that are not presented in this study. For future research work, the most beneficial system configuration has to be found with respect to full year operation including transition periods as well. Additionally, other desiccant materials will be investigated. Furthermore, a system simulation model is created in order to carry out dynamic system simulations to analyze applicability of the investigated system for different locations during full year operation as well as different system setups.

Author Contributions: Conceptualization, P.N., F.R. and A.S.; methodology, P.N. and A.S.; software, P.N. and A.S.; validation, P.N., F.R. and A.S.; formal analysis, P.N., F.R. and A.S.; investigation, P.N. and A.S.; writing—original draft preparation, P.N.; writing—review and editing, P.N.; visualization, P.N.; supervision, P.N., F.R. and A.S.; project administration, F.R.; G.S.; funding acquisition, G.S. All authors have read and approved the final manuscript.

Funding: This work is being conducted in the frame of a project funded by the Federal Ministry for Economic Affairs and Energy (www.bmwi.de), cf. project funding reference number 03ET1421A.

Conflicts of Interest: The authors declare no conflict of interest.

Nomenclature

Latin symbols

A	area, m^2
c_p	specific heat capacity, $J \cdot kg^{-1} \cdot K^{-1}$
COP	coefficient of performance, dimensionless value
d	diameter, m
DCOP	dehumidification efficiency, dimensionless value
h	specific enthalpy, $J \cdot kg^{-1}$
$\dot{m}$	mass flow, $kg \cdot s^{-1}$
MPF	monthly performance factor, dimensionless value
P	electrical power, W
Q	thermal energy, J
$\dot{Q}$	thermal power, W
r_0	specific evaporation enthalpy, $J \cdot kg^{-1}$
SPF	seasonal performance factor, dimensionless value
$\dot{V}$	volume flow, $m^3 \cdot s^{-1}$
W	electrical energy, J
x	water content, $g_w\, kg_{air}^{-1}$

Greek symbols

ϑ	temperature, °C
λ	thermal conductivity, $W \cdot m^{-1} \cdot K^{-1}$
τ	time, s
φ	relative humidity,% rh
Φ	heat recovery efficiency, dimensionless value
Ψ	moisture recovery efficiency, dimensionless value

Dimensionless values

Δ	difference
$\overline{[\cdots]}$	averaged quantity, dimensionless value

Subscripts and Abbreviations

AH	air heater
AH-GEO	"**A**diabatic **H**umidification and **GEO**thermal system", reference system with adiabatic humidification and heat pump (winter)
AHU	air handling unit
AUX	auxiliary energies
BHE	borehole heat exchanger
CHP	combined heat and power generation
DP-VC	"**D**ew **P**oint and **V**apor **C**ompression chiller", reference system with vapor compression chiller and dehumidification by cooling below dew point temperature (summer)
DW	desiccant wheel
DW-GEO	"**D**esiccant **W**heel and **GEO**thermal system", investigated system
DW-VC	"**D**esiccant **W**heel and **V**apor **C**ompression chiller", reference system with vapor compression chiller and desiccant assisted dehumidification (summer)
el	electrical
eta	extract air
exa	exhaust air
EW	enthalpy wheel
GCHP	ground-coupled heat pump
h	heating
HRW	heat recovery wheel
HVAC	heating, ventilation and air conditioning
IH-GEO	"**I**sothermal **H**umidification and **GEO**thermal system", reference system with isothermal humidification and heat pump (winter)
in	inlet

m	month
max	maximum
min	minimum
nom	nominal
oda	outside air
out	outlet
p	period
PU	circulation pump
Ref	reference
RH	reheater
set	set point
STU	solar thermal unit
su	summer
sup	supply air
th	thermal
UHS	underfloor heating system
w	water
wi	winter

References

1. Davis, L.W.; Gertler, P.J. Contribution of air conditioning adoption to future energy use under global warming. *Proc. Natl. Acad. Sci. USA* **2015**, *112*, 5962–5967. [CrossRef] [PubMed]
2. International Energy Agency. The Future of Cooling—Opportunities for Energy-Efficient Air Conditioning. Available online: https://webstore.iea.org/the-future-of-cooling (accessed on 20 March 2019).
3. International Energy Agency. Energy Technology Perspectives 2010—Scenarios & Strategies to 2050. Available online: https://webstore.iea.org/energy-technology-perspectives-2010 (accessed on 20 March 2019).
4. Isaac, M.; Van Vuuren, D.P. Modeling global residential sector energy demand for heating and air conditioning in the context of climate change. *Energy Policy* **2009**, *37*, 507–521. [CrossRef]
5. Enteria, N.; Mizutani, K. The role of the thermally activated desiccant cooling technologies in the issue of energy and environment. *Renew. Sustain. Energy Rev.* **2011**, *15*, 2095–2122. [CrossRef]
6. Vakiloroaya, V.; Samali, B.; Fakhar, A.; Pishghadam, K. A review of different strategies for HVAC energy saving. *Energy Convers. Manag.* **2014**, *77*, 738–754. [CrossRef]
7. Shehadi, M. Review on humidity control technologies in buildings. *J. Build. Eng.* **2018**, *19*, 539–551. [CrossRef]
8. Jani, D.B.; Mishra, M.; Sahoo, P.K. Solid desiccant air conditioning—A state of the art review. *Renew. Sustain. Energy Rev.* **2016**, *60*, 1451–1469. [CrossRef]
9. Kojok, F.; Fardoun, F.; Younes, R.; Outbib, R. Hybrid cooling systems: A review and an optimized selection scheme. *Renew. Sustain. Energy Rev.* **2016**, *65*, 57–80. [CrossRef]
10. Ge, T.S.; Dai, Y.J.; Wang, R.Z. Review on solar powered rotary desiccant wheel cooling system. *Renew. Sustain. Energy Rev.* **2014**, *39*, 476–497. [CrossRef]
11. La, D.; Dai, Y.J.; Li, Y.; Wang, R.Z.; Ge, T.S. Technical development of rotary desiccant dehumidification and air conditioning: A review. *Renew. Sustain. Energy Rev.* **2010**, *14*, 130–147. [CrossRef]
12. Al-Alili, A.; Hwang, Y.; Radermacher, R. Performance of a desiccant wheel cycle utilizing new zeolite material: Experimental investigation. *Energy* **2015**, *81*, 137–145. [CrossRef]
13. Eicker, U.; Schürger, U.; Köhler, M.; Ge, T.; Dai, Y.; Li, H.; Wang, R. Experimental investigations on desiccant wheels. *Appl. Therm. Eng.* **2012**, *42*, 71–80. [CrossRef]
14. Ruivo, C.R.; Angrisani, G.; Minichiello, F. Influence of the rotational speed on the effectiveness parameters of a desiccant wheel: An assessment using experimental data and manufacturer software. *Renew. Energy* **2015**, *76*, 484–493. [CrossRef]
15. Angrisani, G.; Minichiello, F.; Sasso, M. Improvements of an unconventional desiccant air conditioning system based on experimental investigations. *Energy Convers. Manag.* **2016**, *112*, 423–434. [CrossRef]
16. Chen, C.-H.; Hsu, C.-Y.; Chen, C.-C.; Chiang, Y.-C.; Chen, S.-L. Silica gel/polymer composite desiccant wheel combined with heat pump for air-conditioning systems. *Energy* **2016**, *94*, 87–99. [CrossRef]

17. Angrisani, G.; Roselli, C.; Sasso, M. Experimental assessment of the energy performance of a hybrid desiccant cooling system and comparison with other air-conditioning technologies. *Appl. Energy* **2015**, *138*, 533–545. [CrossRef]
18. Al-Alili, A.; Hwang, Y.; Radermacher, R. A hybrid solar air conditioner: Experimental investigation. *Int. J. Refrig.* **2014**, *39*, 117–124. [CrossRef]
19. Fong, K.F.; Lee, C.K.; Chow, T.T.; Fong, A.M.L. Investigation on solar hybrid desiccant cooling system for commercial premises with high latent cooling load in subtropical Hong Kong. *Appl. Therm. Eng.* **2011**, *31*, 3393–3401. [CrossRef]
20. Ghali, K. Energy savings potential of a hybrid desiccant dehumidification air conditioning system in Beirut. *Energy Convers. Manag.* **2008**, *49*, 3387–3390. [CrossRef]
21. Mazzei, P.; Minichiello, F.; Palma, D. Desiccant HVAC systems for commercial buildings. *Appl. Therm. Eng.* **2002**, *22*, 545–560. [CrossRef]
22. El-Agouz, S.A.; Kabeel, A.E. Performance of air conditioning system with geothermal energy under different climatic conditions. *Energy Convers. Manag.* **2014**, *88*, 464–475. [CrossRef]
23. Wrobel, J.; Walter, P.S.; Schmitz, G. Performance of a solar assisted air conditioning system at different locations. *Sol. Energy* **2013**, *92*, 69–83. [CrossRef]
24. Casas, W.; Schmitz, G. Experiences with a gas driven, desiccant assisted air conditioning system with geothermal energy for an office building. *Energy Build.* **2005**, *37*, 493–501. [CrossRef]
25. Angrisani, J.; Diglio, G.; Sasso, M.; Calise, F.; Dentice d'Accadia, M. Design of novel geothermal heating and cooling system: Energy and economic analysis. *Energy Convers. Manag.* **2016**, *108*, 144–159. [CrossRef]
26. Beccali, M.; Finocchiaro, P.; Nocke, B. Energy performance of a demo solar desiccant cooling system with heat recovery for the regeneration of the adsorption material. *Renew. Energy* **2012**, *44*, 40–52. [CrossRef]
27. De Antonellis, S.; Intini, M.; Joppolo, C.M.; Molinaroli, L.; Romano, F. Desiccant wheels for air humidification: An experimental and numerical analysis. *Energy Convers. Manag.* **2015**, *106*, 355–364. [CrossRef]
28. Kawamoto, K.; Cho, W.; Kohno, H.; Koganei, M.; Ooka, R.; Kato, S. Field Study on Humidification Performance of a Desiccant Air-Conditioning System Combined with a Heat Pump. *Energies* **2016**, *9*, 89. [CrossRef]
29. La, D.; Dai, Y.; Li, H.; Li, Y.; Kiplagat, J.K.; Wang, R. Experimental investigation and theoretical analysis of solar heating and humidification system with desiccant rotor. *Energy Build.* **2011**, *43*, 1113–1122. [CrossRef]
30. Preisler, A.; Brychta, M. High potential of full year operation with solar driven desiccant evaporative cooling systems. *Energy Procedia* **2012**, *30*, 668–675. [CrossRef]
31. Zhang, L.Z.; Niu, J.L. Performance comparisons of desiccant wheels for air dehumidification and enthalpy recovery. *Appl. Therm. Eng.* **2002**, *22*, 1347–1367. [CrossRef]
32. Strindehag, O.; Josefsson, I.; Henningson, E. Emission of bacteria from air humidifiers. *Environ. Int.* **1991**, *17*, 235–241. [CrossRef]
33. Speerforck, A.; Schmitz, G. Experimental investigation of a ground-coupled desiccant assisted air conditioning system. *Appl. Energy* **2016**, *181*, 575–585. [CrossRef]
34. Speerforck, A. *Investigation of a Desiccant Assisted Geothermal Air Conditioning System*, 1st ed.; Verlag Dr. Hut: Munich, Germany, 2019.
35. Speerforck, A.; Ling, J.; Aute, V.; Radermacher, R.; Schmitz, G. Modeling and simulation of a desiccant assisted solar and geothermal air conditioning system. *Energy* **2017**, *141*, 2321–2336. [CrossRef]
36. Niemann, P.; Richter, F.; Speerforck, A.; Schmitz, G. Performance investigation of a desiccant assisted solar and geothermal air conditioning system during winter and summer. In Proceedings of the International Sustainable Energy Conference on Renewable Heating and Cooling in Integrated Urban and Industrial Energy Systems, Graz, Austria, 3–5 October 2018; pp. 208–217. [CrossRef]
37. Slayzak, S.J.; Ryan, J.P. *Desiccant Dehumidification Wheel Test Guide: Technical Report*; No. NREL/TP-550-26131; National Renewable Energy Laboratory: Golden, CO, USA, 2000.
38. *Indoor Environmental Input for Design and Assessment of Energy Performance of Buildings Addressing Indoor Air Quality, Thermal Environment, Lightning and Acoustics*, 12-2012 ed.; DIN EN 15251; Beuth: Berlin, Germany, 2012.

MDPI

St. Alban-Anlage 66

4052 Basel

Switzerland

Tel. +41 61 683 77 34

Fax +41 61 302 89 18

www.mdpi.com

Energies Editorial Office

E-mail: energies@mdpi.com

www.mdpi.com/journal/energies